T0383043

System of Systems Modeling and Analysis

This book provides the reader with motivation, theory, methodology, and examples of modeling and analysis for system of system (SoS) problems. In addition to theory, the text contains history and conceptual definitions, as well as the theoretical fundamentals of SoS modeling and analysis. It then describes methods for SoS modeling and analysis, including use of existing methodology and original work, specifically oriented to SoS.

Providing a bridge between theory and practice for modeling and analysis of SoS, *System of Systems Modeling and Analysis* includes generalized concepts and Methods, Tools, and Processes (MTP) applicable to SoS across any application domain. Examples of application from various fields will be used to provide a practical demonstration of the use of the methodologies.

Features

- Offers a modern presentation of SoS principles and guided description of applying a modeling and analysis process to SoS engineering
- Provides additional modeling approaches useful for SoS engineering, including agent-based modeling
- Covers the current gap in literature between theory and modeling/application
- Features examples of applications from various fields, such as energy grids and regional transportation
- Includes questions, examples, and exercises at the end of each chapter

This book is intended for senior undergraduate students in engineering programs studying SoS modeling, SoS analysis, and SoS engineering courses. Professional engineers will also benefit from MTP and examples as a baseline for specific user applications.

SYSTEM OF SYSTEMS ENGINEERING

Series Editor

Mo Jamshidi

System of Systems Engineering: Principles and Applications
Edited by Mo Jamshidi

Discrete-Time Inverse Optimal Control for Nonlinear Systems
Edgar N. Sanches, Gernando Ornelas- Tellez

Netcentric System of Systems Engineering with DEVS Unified Process
Saurabh Mittal, Jose L. Risco Martin

Intelligent Control Systems with an Introduction to System of Systems Engineering
Thrishantha Nanayakkara, Ferat Sahin, Mo Jamshidi

Emergent Behavior in System of Systems Engineering: Real World Applications
Larry B. Rainey, Thomas O. Holland

System of Systems Modeling and Analysis
Daniel A. DeLaurentis, Kushal Moolchandani, Cesare Guariniello

For more information about this series, please visit: https://www.routledge.com/System-of-Systems-Engineering/book-series/CRCSYSENGELE

System of Systems Modeling and Analysis

Daniel A. DeLaurentis
Kushal Moolchandani
Cesare Guariniello

CRC Press
Taylor & Francis Group
Boca Raton London New York

CRC Press is an imprint of the
Taylor & Francis Group, an **informa** business

MATLAB® is a trademark of The MathWorks, Inc. and is used with permission. The MathWorks does not warrant the accuracy of the text or exercises in this book. This book's use or discussion of MATLAB® software or related products does not constitute endorsement or sponsorship by The MathWorks of a particular pedagogical approach or particular use of the MATLAB® software.

First edition published 2023
by CRC Press
6000 Broken Sound Parkway NW, Suite 300, Boca Raton, FL 33487-2742

and by CRC Press
4 Park Square, Milton Park, Abingdon, Oxon, OX14 4RN

CRC Press is an imprint of Taylor & Francis Group, LLC

ISBN: 978-1-032-13830-5 (hbk)
ISBN: 978-1-032-13836-7 (pbk)
ISBN: 978-1-003-23101-1 (ebk)

DOI: 10.1201/9781003231011

Typeset in Nimbus Roman
by KnowledgeWorks Global Ltd.

Dedication

DD *To Prof. Mo Jamshidi, dedicated builder of System of Systems community; to my early mentors: Daniel P. Schrage, Dimitri N. Mavris, and to SoS generational thinker Robert K. Callaway*

KM *To papa, mummy, and Harsha; and to all others who helped me reach where I am today*

CG *To my mom and dad, my sister, and all the friends and relatives who never ceased to support my desire to learn*

Contents

Preface......xv

Contributors......xvii

Authors......xix

PART I Systems of Systems: Theory and Process for Modeling and Analysis

Chapter 1 What Is a System of Systems?......3

1.1 Motivation: A Better Approach for Generational Challenges......3
1.2 Systems and Systems Thinking......5
1.3 Defining *Systems*......8
1.3.1 Complex Systems......9
1.4 Brief History of System of Systems......11
1.5 What is a System of Systems?......13
1.5.1 Distinguishing SoS on Attributes......14
1.5.2 Classifying SoS on Control Authority......16
1.5.2.1 Directed System of Systems......16
1.5.2.2 Acknowledged System of Systems......17
1.5.2.3 Collaborative System of Systems......18
1.5.2.4 Virtual System of Systems......19
1.6 Chapter Summary......19
1.7 Discussion Questions and Exercises......21
1.7.1 Discussion Questions......21
1.7.2 Exercises......21

Chapter 2 What Is System of Systems Engineering?......23

2.1 Overview of *Systems Engineering*......23
2.2 Role of Modeling and Analysis......26
2.3 System of Systems Engineering......29
2.3.1 What is Different?......29
2.3.2 Systems Thinking: A Key to SoSE......31

2.3.3 Distinctive Features of SoSE 33
2.3.4 SoSE: An Industrial Snapshot....34
2.3.5 SoSE in Academia: Small Start, Steady Growth 35
2.4 Example: Air Transportation System as a System of Systems 36
2.5 Chapter Summary 36
2.6 Discussion Questions....37

Chapter 3 A Formal Process of SoS Modeling and Analysis 39

3.1 SoS Representation....39
3.1.1 SoS Representation and Hierarchy 39
3.1.2 Lexicon 41
3.1.3 Taxonomy 43
3.2 A 3-Phase Method for SoS Problems 45
3.3 Definition Phase and Tools....45
3.3.1 Example Definition: ATS 48
3.4 Abstraction Phase and Tools....49
3.4.1 Example Abstraction: ATS 52
3.5 Implementation Phase and Tools....53
3.5.1 Verification and Validation....54
3.5.2 Example Implementation: ATS....56
3.6 Chapter Summary 56
3.7 Discussion Questions and Exercises....58
3.7.1 Discussion Questions....58
3.7.2 Exercises 59

PART II Methods and Tools for System of Systems Modeling and Analysis

Chapter 4 Bridging Theory and Practice....63

4.1 Choosing the Right Questions 63
4.2 Choosing the Right Tools 65
4.3 Putting it All Together: The SoS M&A Project....76
4.3.1 Overall Project Description....77
4.3.1.1 Overview 77
4.3.1.2 Project Learning Objectives 77
4.3.1.3 Project Technical Objective....77
4.3.1.4 Project Deliverable Items 78
4.3.2 Definition Phase Deliverable 78
4.3.2.1 Format....78
4.3.2.2 Procedure....78

4.3.3 Abstraction Phase Deliverable 79
4.3.3.1 Format 79
4.3.3.2 Procedure 79
4.3.4 Implementation Phase Deliverable 80
4.3.4.1 Deliverable Items and Expectations 80
4.3.4.2 Report Specifics 81
4.3.4.3 Poster Presentation Specifics 81
4.4 Notes on Applications and Selected List of SoS M&A Project Topics 82
4.5 Chapter Summary 84

Chapter 5 Network Theory 85

5.1 Basic Graph Theory and Network Measures 85
5.1.1 Types of networks 86
5.1.1.1 Undirected Network 86
5.1.1.2 Directed Network 86
5.1.1.3 Hypergraphs 87
5.1.1.4 Bipartite Graphs 87
5.1.1.5 Network Types Based on Topology 88
5.1.2 Measures in Network Theory 89
5.1.2.1 Degree, Degree Distribution, and Network Density 90
5.1.2.2 Paths, Cycles, and Network Diameter 92
5.1.2.3 Clustering Coefficient 93
5.1.2.4 Node Centrality 94
5.1.2.5 Assortativity 95
5.2 Modeling Network Dynamics 95
5.2.1 Growth Algorithms: Random Network 97
5.2.2 Growth Algorithms: Scale-Free Network 97
5.3 Using Networks for SoS Modeling and Analysis 98
5.3.1 Modeling Interactions and Flow within Networks 99
5.3.2 Behaviors of Complex Networks 101
5.3.2.1 Example 4: Disease Spread via Random Geometric Graphs 101
5.4 Modeling the Air Transportation System Using Network Theory 102
5.5 Chapter Summary 104
5.6 Discussion Questions and Exercises 105
5.6.1 Discussion Questions 105
5.6.2 Exercises 105

Chapter 6 Agent-Based Modeling 107

6.1 A Brief Introduction to Agent-Based Modeling 107
6.1.1 What are Agents? 110

6.1.2 Types of Agents ... 111
6.2 The When and Why of ABM ... 112
6.3 Agent-Based Modeling for System of Systems ... 115
6.4 Examples of Application to System of Systems ... 117
6.5 Chapter Summary ... 118
6.6 Discussion Questions and Exercises ... 119
6.6.1 Discussion Questions ... 119
6.6.2 Exercises ... 120

Chapter 7 Specialized Methods and Tools for System of Systems Engineering ... 121

7.1 Analytic Workbench ... 121
7.1.1 Robust Portfolio Optimization ... 123
7.1.1.1 General Formulation of Investment Portfolio Approach ... 123
7.1.1.2 Robust Mean Variance Optimization ... 125
7.1.1.3 The Bertsimas-Sim Method ... 129
7.1.1.4 Conditional Value-at-Risk Optimization ... 130
7.1.2 Systems Operational Dependency Analysis ... 132
7.1.2.1 Genesis of the Model ... 132
7.1.2.2 Operational Dependencies ... 133
7.1.2.3 Model Parameters ... 134
7.1.2.4 Modeling Dependency on a Single System ... 136
7.1.2.5 Modeling Dependency on Multiple Systems ... 137
7.1.2.6 Evolution of Operability over Time and Robustness ... 139
7.1.2.7 Robustness and Resilience ... 139
7.1.2.8 Deterministic Analysis ... 142
7.1.2.9 Stochastic Analysis ... 143
7.1.2.10 Synthesis and Architectural Design Updates ... 144
7.1.2.11 SODA Problem Setup ... 144
7.1.2.12 Source of Parameters ... 149
7.1.3 Example of Application of SODA: an Earth Observation System ... 151
7.1.3.1 Deterministic Analysis ... 152
7.1.3.2 Stochastic Analysis ... 153
7.1.3.3 Flexibility and Resilience ... 155
7.1.4 Systems Developmental Dependency Analysis ... 157
7.1.4.1 Developmental Dependencies ... 159
7.1.4.2 Parameters of the Model ... 160

7.1.4.3 Basic Formulation of SDDA 161
7.1.4.4 Conservative Formulation of SDDA 163
7.1.4.5 Deterministic Analysis 163
7.1.4.6 Stochastic Analysis................................ 166
7.1.4.7 Source of Parameters 169
7.1.5 Example of Application of SDDA: a Communication 170
7.1.5.1 Delay Absorption.................................. 172
7.1.6 Multi-stakeholder Dynamic Optimization 175
7.1.6.1 SoS Manager's Problem 176
7.1.6.2 SoS Participant's Problem 177
7.1.6.3 Transfer Contract Coordination Mechanism and Approximate Dynamic Programming 178
7.2 Other Useful Methods for SoS Modeling and Analysis 180
7.2.1 System Dynamics... 180
7.2.2 Design Structure Matrix.. 182
7.3 Chapter Summary 182

Chapter 8 Enhancing System of Systems Engineering 185

8.1 Artificial Intelligence, Machine Learning, and Autonomy 185
8.1.1 AI/ML as Driver and Analyzer 186
8.1.2 AI/ML for Extraction and Analysis of Data 188
8.2 Uncertainty 190
8.2.1 Uncertainty in System of Systems 190
8.2.2 Uncertainty Quantification.................................... 192
8.3 Complexity 193
8.3.1 Can Complexity Aid SoS M&A? 193
8.3.2 Effective Complexity and Complex Adaptive Systems 194
8.3.3 Sources of Complexity.. 196
8.3.4 Complexity Metrics 197
8.3.5 Example of an SoS-relevant Complexity Metric ... 199
8.4 Model-Based Systems of Systems Engineering 199
8.5 Chapter Summary 201

PART III Examples of Application of System of Systems Modeling and Analysis

Chapter 9 Advanced Air Transportation System of Systems....................... 205

9.1 Problem Introduction.. 205
9.2 Definition Phase... 206
9.2.1 Operational Context, Status Quo, and Barriers...... 206

9.2.2 Scope Categories and Levels 208
9.2.2.1 Example 1: Robust, Scalable Transportation System Concept 209
9.2.2.2 Example 2: Assessing New Technologies on Future Fleet and Emissions 210
9.2.2.3 Example 3: Air Transportation Network Restructuring 210
9.3 Abstraction Phase 211
9.3.1 Resources, Stakeholders, and Networks 211
9.3.2 Drivers and Disruptors 212
9.3.2.1 Example 1: Robust, Scalable Transportation System Concept 212
9.3.2.2 Example 2: Assessing New Technologies on Future Fleet and Emissions 213
9.3.2.3 Example 3: Air Transportation Network Restructuring 213
9.4 Implementation Phase 215
9.4.1 Example 1: Robust, Scalable Transportation System Concept 217
9.4.2 Example 2: Assessing New Technologies on Future Fleet and Emissions 218
9.4.3 Example 3: Air Transportation Network Restructuring 218

Chapter 10 Human Space Exploration System of Systems 221

10.1 Problem Introduction 221
10.2 Definition Phase 222
10.2.1 Hierarchy of the Mars Exploration Architectures 222
10.3 Abstraction Phase 228
10.3.1 γ Level 229
10.3.2 β Level 229
10.3.3 α Level 230
10.4 Implementation Phase 232
10.4.1 "What-ifs" and SODA Analysis 234
10.4.2 "What-ifs" and SDDA Analysis 235
10.4.3 "What-ifs" and Combined SODA/SDDA Analysis 238
10.5 Beyond the Initial SoS 239
10.5.1 Lower Level of Abstraction: Propulsion Systems and Lunar Gateway Habitat Subsystems 240

10.5.1.1 Propulsion Systems 241
10.5.1.2 Lunar Gateway Habitat 244
10.5.2 Including other Aspects of ROPE: Budget and Policies for Technology Prioritization 245

Glossary 253

References 255

Index 265

Preface

A System of Systems (SoS) is a particular class of systems whose constituent elements are themselves independently operable and useful systems. However, by working together as a network of interacting systems, the collective whole can produce capabilities that none of them could do alone. While a similar statement can be made for systems, some differences in the case of SoS are that the boundaries of an SoS are usually more fluid and dependent on the stakeholder defining them, their constituent systems cannot only be operated, but also managed independently, and the constituent systems frequently have their own objectives for participating in the collective which may be different from those of the entire SoS.

Given the difficulty of narrowing down to a single definition, it is useful to identify characteristics that allow for categorization of systems as SoS than having a strict definition of the term. But the lack of a definition is not the only shortcoming. Due to the complexity of SoS, the tools and methods utilized for engineering them are equally diverse and complex. As a result, the field of SoS engineering (SoSE) offers many exciting opportunities for the advancement of both theory and practice.

In this book, we present the concept of an SoS, its distinguishing traits, and the lexicon and taxonomy used in describing it. We include the theory for classification and description of SoS and demonstrate all theory with examples of application to air transportation and space systems engineering problems. However, because this book started as a set of lecture notes for the *AAE 560: Systems-of-Systems Modeling & Analysis* course taught at Purdue University's School of Aeronautics and Astronautics, it is dedicated in large part to the modeling and analysis of SoS. It is also dedicated in a different sense, to many students (including a large cadre of distance learning students enrolled in the course while simultaneously facing the challenges of SoS in their jobs) who truly enhanced the development of this material and the components herein through their hard work, thoughtful questions and conversations, and tackling many assessments, homework assignments, group projects, and monster midterm exams! Finally, and without question, the ideas, concepts, methods, and insights in this book can be attributed to all the present and former members of the System of Systems Lab (SoSL) and Center for Integrated Systems in Aerospace (CISA) at Purdue. To each of these groups, THANK YOU.

We knew from the start that such a course, delivered largely (but not exclusively) to engineering students, would need to guard against the eternal temptation to "start coding" and modeling before the problem to be solved was sufficiently defined. Thus, we ask that you heed the sage advice of famous mathematician Richard Bellman that, "The right problem is always so much harder than a good solution." In the world of SoS, this is so very important.

We present a three-phase approach for modeling and analysis (M&A) of SoS problems. Through the three phases, we define a problem, abstract and represent the system, and finally demonstrate approaches for generating solutions. For the purpose of problem representation and solution, we make use of modeling and simulation approaches from network science and agent-based modeling. The chapters on modeling methods are intended to present a brief introduction to their respective subjects, focusing only on the material necessary to understand their application to SoS problems discussed in the book. Additional detailed information on these topics can be found in the related bibliography.

Research on SoS and their problems of design and analysis is ongoing. In the final part of the book, we illustrate the application of SoS modeling and analysis methods to an advanced air mobility problem and a space systems engineering problem and present some open research questions that still need to be addressed for the advancement of SoS theory. We hope this discussion will serve to excite interest in advancing the theory and practice of SoS.

Overall, we have attempted the ever-dangerous task of producing a book useful for several audiences! However, our primary audience is university students (and their instructors). The focus on fundamentals, ample references, and rich set of discussion/exercise questions provide, we hope, a solid basis for a senior undergraduate and/or master's level course. Our confidence in serving this audience stems from the fact that this book is largely derived from fifteen years of wonderful experiences from the AAE 560 course. A secondary audience consists of researchers (whether in academia, industry, or government settings), where our goal is assisting them by connecting many theories and threads of scientific and engineering innovation to the evolving challenge of SoS M&A. In addition, motivating the right research mindset is a key message to tackle the many challenges that remain in this area. Finally, for SoS-oriented practitioners, we hope that the mindset of identifying the right problem, along with the structured SoS M&A process to address it, will enhance their execution of their job functions. Especially for these latter two audiences we aspire (using this term because we will never attain!) to the style and effect of Herbert Simon's *The Sciences of the Artificial* [149], a book that has very much shaped our research, teaching, and certainly portions of this book. Simon's book dealt with deep technical issues, but in an eminently readable manner and compelling style.

Contributors

Prajwal Balasubramani
Purdue University
West Lafayette, Indiana

Derek Carpenter
Purdue University
West Lafayette, Indiana

Liam Durbin
Purdue University
West Lafayette, Indiana

Kshitij Mall
Purdue University
West Lafayette, Indiana

Authors

Daniel A. DeLaurentis, PhD, is professor in Purdue University's School of Aeronautics and Astronautics, where he also directs the Center for Integrated Systems in Aerospace (CISA) and its main component, the System of Systems Lab. His primary research and teaching interests include problem formulation, modeling, design and control methods for aerospace systems and systems-of-systems, all from a model-based perspective. Dr. DeLaurentis annually teaches a graduate course "System of Systems Modeling and Analysis" at Purdue, now regularly with enrollment over 100 students. DeLaurentis has supported many students and professional staff who have made impacts across diverse domains including air transportation, defense/security, civil infrastructure, and space exploration. Dr. DeLaurentis also as the Chief Scientist of the U.S. DoD's Systems Engineering Research Center (SERC). He is Fellow of the International Council on Systems Engineering (INCOSE), Associate Fellow of the American Institute of Aeronautics and Astronautics (AIAA), and senior member of the IEEE.

Kushal Moolchandani, PhD, works as an aerospace engineer for Universities Space Research Association at the NASA Ames Research Center. His current work is research under NASA's Air Traffic Management - eXploration (ATM-X) project, specifically on the development of airspace services for Urban Air Mobility. Previously, he has also worked on design and optimization of civil transport aircraft and modeling and simulation studies of aviation's impact on the environment. His research interests include design and optimization of complex aerospace systems, air transportation systems, and systems engineering. He earned his Bachelor of Engineering in Aeronautical Engineering from PEC University of Technology, Chandigarh, India, and MS and PhD in Aeronautics and Astronautics from Purdue University.

Cesare Guariniello, PhD, is a research scientist in Purdue University's School of Aeronautics and Astronautics. He holds two Master's degrees, in Automation and Robotics Engineering and in Astronautical Engineering, from the University of Rome "La Sapienza", and a PhD in Aeronautics and Astronautics from Purdue University. His research interests include System-of-Systems design and architecting, space applications, cybersecurity, dynamics and control, and planetary geology. Dr. Guariniello has been involved in research projects with NASA, the US DoD, the US Navy, MITRE corporation, and the NSF. Recently, he expanded his role to mentoring and managing students and lecturing graduate-level classes. Dr. Guariniello is a senior member of the American Institute of Aeronautics and Astronautics (AIAA) and the Institute of Electrical and Electronical Engineers (IEEE), and a member of the International Council on Systems Engineering (INCOSE), and the American Astronautical Society (AAS).

Part I

Systems of Systems: Theory and Process for Modeling and Analysis

1 What Is a System of Systems?

1.1 MOTIVATION: A BETTER APPROACH FOR GENERATIONAL CHALLENGES

Those of us with a passion for designing the next-generation 'thing' are always doing a good bit of forecasting and dreaming about the future. For us authors, our particular passion began in aerospace, motivating us to ask how we could create a new aerospace vehicle, equipped with a few yet-to-be-realized technologies, with the highest probability of success as perceived by the most interested stakeholders. It is this kind of desire, applied to challenges far broader than aerospace vehicles, that motivates the research and teaching we have done and the genesis for this book.

Our journey had many starting points. Prime among them was a collection of conversations circa 2003-2004 among related thinkers and kindred spirits sharing an eagerness to think holistically and long-term. Part of this conversation was spurred by a compelling study report from the RAND Corporation titled, "Shaping the Next One Hundred Years: New Methods for Quantitative, Long-Term Policy Analysis" [116]. Authored by Robert Lempert (a physicist turned policy decision scientist), Steven Popper (an economist), and Steven Banks (a computer scientist), this report spoke eloquently about the need to improve our abilities to grapple with generational challenges and then started to deliver on that need by offering a new way to model and analyze in this context for robust decision-making in the area of Long Term Policy Analysis (LTPA). Their approach brought new analysis tools and measures in the critical component areas of generating ensembles of scenarios in an organized way, finding robust strategies open to adaptation (vs. fixed point strategies), and providing means for decision-makers to interact with the computer tools and their outputs. These are now hallmarks of how to approach such challenges and are central to the modeling and analysis approaches for System of Systems (SoS) to be covered in this book. We will return to some of their key concepts in more depth later.

In some sense, the diverse backgrounds of the RAND report authors is as instructive for us as the content of the report. The generational challenges we are all interested in cannot be "engineered" in the traditional sense that we engineers would engineer a compressor, a turbine engine, or even an entire aircraft. From the earliest days of the journey, we and fellow sojourners have had to focus intently on maintaining the mantra of "trans-domain" thinking, and modeling, rather than mere multidisciplinary or worse yet "siloed".

Also, around this 2003-2004 time frame and in the few years preceding, there had been much thinking about the future of transportation. Note: There are *always* people thinking about the future of transportation and that is a good thing! At that particular time, innovation was flourishing on the air side with the NASA Small Aircraft Transportation System (SATS) program looking to unlock the possibilities via thousands of underutilized local and regional airports as well as emerging work on Personal Air Vehicles (PAVs) and a bit later on proto-air-taxi concepts. But there was also strong push for change in surface transport innovations including high speed rail. How does one reason about all of these options, in the midst of "deep uncertainty" (as the RAND team would put it), to include far more than just technological uncertainty, but policy and economics as well?

Yet in the midst of this conundrum – which we likened to navigating a minefield without a map – was awareness of the tremendous opportunity to influence long-term consequences by making more enlightened decisions now. But, clearly better tools were needed. This was the message of author DeLaurentis and early collaborator Robert Callaway in their 2004 article making the case for SoS as the key concept to generate that needed map [46]. Resonating with the emerging idea that thinking "bigger" than just the aircraft (or airport, or train) was essential, they presented constructs to reason about the systematic integration of all these possible systems, technologies, policies and economic drivers... yes, a "System of Systems"! Recognizing that SoS was in fact the "system of interest" was an important aspect; but so was the idea that, rather than eliminate the silos, we needed a few folks brave enough to span them and approach generational challenges via SoS solutions. It was also clear that these brave people must be ready to continually test hypotheses around proper boundary, be humble in the face of deep uncertainty, and be patient in defining the problem in a trans-domain fashion.

But why all this talk of "bravery" and a new cadre of solution-makers? We have engineering! The formalized process of "engineering" involves the use of scientific principles, processes, and tools for building systems. For many centuries, humans have been designing tools to create solutions that meet their needs. Early tools like pulleys and levers, which were intended to meet needs such as hunting and lifting, are simple enough that they can be designed and assembled by individuals. These tools are examples of systems whose inputs and outputs share deterministic relationships, modeled using well-known laws of mechanics. Such simple tools, in turn, have been employed to build more sophisticated systems including those which display non-linear behaviors. Such systems are more complex and their design and development is a task undertaken by larger organizations comprised of numerous functions organized into multiple specialized discipline-specific departments. Simple input-output relationships are insufficient for building these systems, with the result that organizations discovered (and sometimes are still discovering) the need for "systems thinking", the formalization of which is the branch of engineering called "systems engineering."

Over time, we have continued to build increasingly more complex systems and now advanced to a stage of multilevel hierarchies of large-scale systems assembled from components which are themselves sophisticated systems. Interactions within and among levels produce an evermore diverse set of capabilities but also increased propensity for non-linear behaviors sensitive to operational context where small changes in input may lead to large and unpredictable variation in output. In this setting, new systems that we build today are expected to work alongside other new and legacy systems in a network influenced by multiple independent operators. We desire, therefore, and quite urgently need to design robust architectures and appropriate interfaces between large numbers of independent systems to obtain ever-increasing capabilities with at least manageable "side effects". Thus we arrive at this special class of large-scale systems called "Systems of Systems."

But SoS is more than just a class of problems. In our experience, it is also an *aspiration!* As vexing as the term may be for some, the notion of an enduring, effective systematic integration of components, a true "system" of systems (vs. a mere amalgamation), is indeed a worthy aspiration. SoS as a class of problems speaks to "the What"; SoS as an aspiration speaks to "the Why". We will explore both throughout this book.

In the remainder of this chapter, we introduce key foundations: the concepts of systems and systems thinking. We use systems thinking for problem solving when we develop models of the world assuming that the world is composed of a set of interconnected elements, and that its behavior can be explained by observing patterns of interactions among its elements. We explain why systems thinking is essential especially for larger systems, classified as complex systems. Finally, we introduce SoS and identify how they can be distinguished from other complex systems.

1.2 SYSTEMS AND SYSTEMS THINKING

Dated as far back as the fourth century BC, the following maxim has been attributed to Aristotle: *"The whole is more than the sum of its parts. The part is more than a fraction of the whole."* This is the notion of thinking about the world in terms of systems, or *systems thinking*, which can be traced back a long time and can be found in the writings of a variety of philosophers including Leibniz, Hegel, Marx, and many more who talked about elements, whole, interactions, and related concepts, albeit with different terminology. For a long time, scientific progress relied on first understanding component elements individually, and then deducing their behavior when they were put together to form a larger whole, i.e., a system. We call this approach *reductionism*, and it was almost the sole voice in science because it was doing well in splitting certain problems into isolated parts with causal relationships. Problems of design or analysis, when they were decomposed into smaller problems, were easier to describe and solve. Consequently, a large body of analytical methods developed for the reductionist approach proved to be successful for solving the specific problems encountered.

Reductionism relies on representing system elements as accurately as possible so that the observed behaviors of the model closely reflect reality. Over time, however, this approach of studying systems in a bottom-up manner starting from its smallest constituent element started to reach the limits of its capability. In fields as diverse as biology, engineering, and social sciences, practitioners started to study much larger and more complex systems, with the result that the mechanistic and causal approach to problem-solving could not continue to do well anymore. We can identify various reasons that reductionist approaches were less useful for larger, more complex systems, including that the sub-systems of these systems themselves begin to behave independently, which the reductionist approaches fail to handle very well. This led to the approach called *holism* which looks at collections of sub-system elements and their interactions as a whole. With a holistic view of the system, model realism is less important than achieving fidelity of representing system behaviors. The result is that we turn our attention from having an accurate system representation to selection of the simplest model which is capable of identifying system behaviors.

The *Guide to the Systems Engineering Body of Knowledge* (SEBoK) [24] traces a similar history of systems science, citing particularly the works of Bertalanffy and Weaver. It describes how during the eighteenth and nineteenth centuries specialized fields of study arose as human knowledge expanded rapidly. However, over time, new problems were encountered which could not be solved by the existing methods. Specifically, the existing methods could not describe the observed emergent behaviors that arise as a consequence of subsystem interactions and a study of a system's isolated parts was not much helpful either. Classical mechanics was useful for studying behaviors of isolated parts, followed by summing these behaviors to describe the wholes. Higher up on the complexity scale, problems of "disorganized complexity" were addressed by statistical mechanics, once again implicitly assuming that it is feasible to study parts of a system in isolation only to later put together a description of the whole. Unfortunately, the classical approaches fall short for problems of "organized complexity," where features such as coordination among and the independence of the parts are paramount.

Thus, starting in the twentieth century, observations of hierarchy and the increase in complexity of systems as we move up the levels of hierarchy started to make an appearance in scientific literature. This naturally led to a search for commonalities in the underlying principles of the various disciplines and the rise of a field which would provide a framework for describing relationships across disciplines and facilitate communication of knowledge across disciplinary boundaries. This was the field of "general systems theory," first described by Bertalanffy [165] and later extended with the idea of hierarchy of levels of system complexity by Boulding [29]. With general systems theory started the search for principles which were applicable to systems from across different disciplines and knowledge-sharing among those disciplines.

Thereafter, there was an explosion in interest in the science of systems and the emergence of fields such as mathematical systems theory, living systems theory, cybernetics, social systems theory, and philosophical systems theory [8]. Cybernetics, which Weiner defined as "the study of control and communication in the animal and the machine," contributed to systems science with the study of how information flows in feedback loops within systems, and the use of this information for control and regulation. Operations research, which arose from problems of military planning during World War II, quickly became an approach grounded in mathematics to support decision-making in organizations. The following decades saw the emergence of fields such as system dynamics and organizational theory, all of which contributed to the advancement of a systems thinking approach to problem solving.

The dramatic scope and scale of the contents of the "Solberg Chart", Figure 1.1, even before each reader lists the dozens of methods missing from this lineage, presents both opportunity and peril. With such a rich collection of amazing "hammers", how do we choose the right one? Well, by fully characterizing the "nail", that is the problem to be solved. This is a central challenge in SoS M&A; the DAI process (Chapter 3) and approaches to establish research questions and hypotheses (Chapter 4) will be our guides to surmount the challenge.

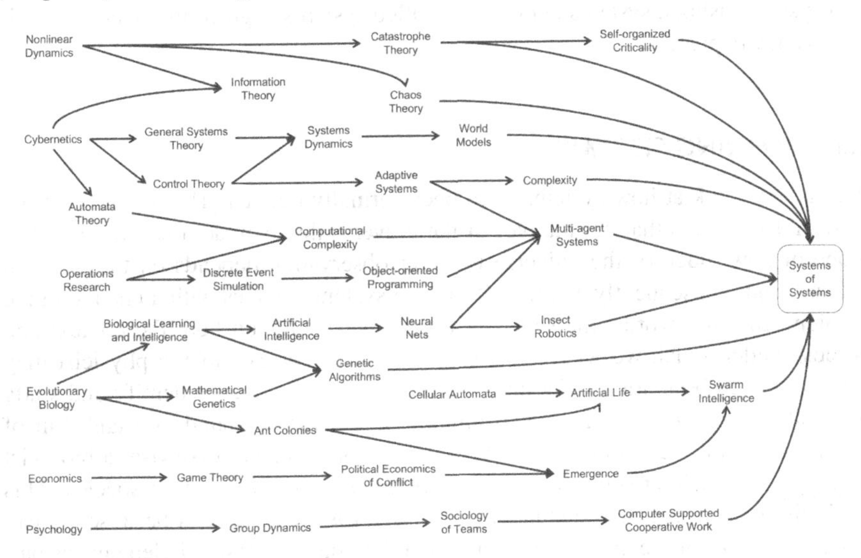

Figure 1.1: The Solberg Chart, a lineage of systems modeling relevant for SoS modeling and analysis.

The totality of what we put forth in this text is founded on the concept of Systems Thinking. *Systems thinking* is the practice of thinking about the world in terms of systems. This involves a decomposition of the structure of the world into a set of distinct constituent elements and considering relationships among constituents to explain how those relationships lead to the observed behaviors of a system. We seek to understand the larger whole from multiple perspectives, rather than analyze behaviors of the constituents in isolation.

The interactions and interdependencies among system components are the reason that the system as a whole provides more capabilities than those provided independently by its parts. These interdependencies are our leverage points, places where changes to design can change the behaviors of the whole. In addition, we regard a system in relation to its environment and evaluate the effects of its operation on those outside its boundaries, and vice versa. We also try to understand the implications of changes, and analogies and parallelism in behaviors of different types of systems [79]. In short, systems thinking involves modeling and analysis of the world using the concepts of a collection of elements, their interconnectedness, and the emergence of behaviors of the whole due to the presence of component interactions.

In summary, systems thinking, or systems approach, is a mode of problem solving wherein attention shifts from parts and the deterministic summation of their effects to an examination of these parts in terms of their role in larger wholes [22]. These wholes are declared to be *systems* whereupon the application of systems thinking becomes useful for tasks such as analysis or prediction of future system behaviors, solving problems, and creating new or modified systems [33]. This last task of design systems using a systems approach is called systems engineering which we will discuss in Chapter 2.

1.3 DEFINING *SYSTEMS*

Let us now look at how systems have been formally defined. The thermodynamic view of a system is that of "the part of the universe that is under observation." The scope of how much of the universe is under observation depends on the problem at hand, and, consequently, the definition of a system changes with a change in the problem. In other words, the scope of a system depends on both the context and purpose under which we wish to study that system. For example, a physicist may want to make observations of a finite portion of the universe to identify and study physical phenomena such as, for example, propagation of sound in a tank full of water, or radiation emitting from distant cosmic phenomena. Likewise, a biologist may study a single creature, a group of creatures, or even an entire species. This flexibility of defining the boundaries of a system is advantageous because we can manage the size of the problem at hand, including only a subset of elements as part of a system that we want to study while leaving out the rest as the *environment*.

Where we set the boundaries of a system depends on what we wish to do with the system. In engineering, our objective is to design systems to fulfill our needs. We do so by iteratively assembling many smaller elements into an architecture so that we obtain the desired collective behaviors. In this case, a possible definition of a system is *a collection of elements that together provide desired functionality.* This, however, is not the sole definition. A *system* is a surprisingly difficult concept to define, as can be seen from the number and diversity of definitions given below:

1. "A system is an assemblage or combination of functionally related elements or parts forming a unitary whole, such as a river system or a transportation system." [22]
2. "A system is a set of elements so interconnected so as to aid in driving toward a defined goal." [74]
3. "A set of different elements connected or related so as to perform a unique function not performable by the elements alone." [122]

4. "A system is an arrangement of parts or elements that together exhibit behavior or meaning that the individual constituents do not." [53]

Three common themes stand out in each one of these definitions, viz., those of a collection of elements, interdependence among these elements, and the presence of certain goals or objectives. *Collection* is related to assembly of elements, meaning that any system is itself composed of multiple subsystems or components; likewise, any system is itself a subsystem in a larger system. *Interdependence* implies that these subsystems affect and are affected by other subsystems that comprise the whole. The nature and structure of these interdependencies influence the behavior of the whole system, hence the notion of 'the whole is greater than the sum of its parts.' Finally, the third theme – *goal* – is particularly relevant for our purposes because it implies that every engineered system is designed to fulfill certain needs; when a system meets our needs we say that it provides a *utility*.

The three features of collection, interdependence, and goals can be present to varying extent in different systems. For example, do natural systems have a goal? As we described, what constitutes a system will depend on what we intend to do with it, and even then not all engineers will agree on the precise boundaries since their subjective worldviews may differ [53]. Moreover, some additional properties may dictate how useful it may be to identify an entity as a system; examples of these properties include emergence, self-regulation to maintain a state or to move towards a desired state, etc. These properties add to the capabilities of a system, but the more capability that a system has, the larger and more complex it tends to become. In fact, over time, it has been a consistent theme that the systems we design have grown larger and their development has become a more expensive undertaking with longer timescales and higher associated risk. Beyond a certain threshold of system size and development effort, we classify such larger systems as *complex systems*, which we discuss next.

1.3.1 COMPLEX SYSTEMS

When discussing the history of systems thinking, we looked at how the reductionist approaches to design started to reach their limit as systems grew larger and more complex. *Complex systems* are a class of systems that usually have a large number of constituent sub-systems. A complex system can be defined as "a system composed of many interacting parts, often called agents, which displays collective behavior that does not follow trivially from the behaviors of the individual parts" [135]. Other similar definitions describe complex systems as those which are "an assembly of interacting members that is difficult to understand as a whole" [11], and systems which are "made by many components interacting in a network structure" [88]. In the former definition, "interacting members" identifies the interdependence of system components, while "difficult to understand as a whole" recognizes that for such systems the system-level behavior is greater than the sum of the behaviors of its components.

Unfortunately, even with these definitions, it is difficult to distinguish systems that are complex from those that are not complex. Instead, we identify those characteristics which make the design and analysis of complex systems difficult and hence warrant a separation classification for such systems. In complex systems, in general, the constituent subsystems are heterogeneous, and their behaviors and the nature of their interactions may change over time. Complex systems are characterized by non-linear dynamics, subsystem interdependencies with feedback loops, and phenomena such as emergence, discussed below. Additionally, complex systems present difficulties in their modeling and prediction of their behaviors, and even in their management [101].

We emphasize the property of emergence because it is of particular significance when distinguishing complex systems [53]. Emergence makes it difficult to predict the behaviors of the whole even when the behaviors of sub-systems are known. Since engineered systems are designed to provide a utility, we wish to model system behaviors to assess them both as sources of the value obtained from the system and a contributor to a system's complexity. Hence, emergence in complex systems may result in counterintuitive outcomes, such as when improvement in the value obtained from sub-systems does not correspond to an equivalent value improvement from the whole system.

From the viewpoint of modeling, even the chosen hierarchy of representation and timescale used may influence if a system is classified as complex. For example, control inputs to a system may have a different short-term response compared to a long-term response. This also contributes to making prediction of emergent behaviors difficult. The system may also be slow to respond to any control input, which would mean the presence of time delays between inputs and outputs of the whole. Choices in representation for modeling are also consequential. In the AAE 560 course, we often start in the very first class meeting by asking the students to describe a simple wooden chair from a systems perspective. After just a few short minutes encircling this obtuse question, there is collective realization that even a simple chair presents many choices. Some offer that, at the highest level of description, it is a mere half-dozen parts: seat, back, legs, a couple of supports, etc. Others describe it in terms of the chemical/material components that make up the wood and any joining material. To the chagrin of all students, the instructor closes the conversation by stating the answer is "it is just a bunch of fermions and bosons!" We will devote much time in the remainder of this book to the topic, the challenge, of effectively representing an SoS as a special kind of complex system.

Clearly, system complexity poses a number of significant challenges to engineers (or problem solvers in any vocation or career field!). Complex systems are also usually difficult to test experimentally. The difficulty in testing may arise in how to conduct the test, or how to manage the cost of the test, or even how to deal with ethical factors of the test. Developing working prototypes may also not be feasible. This is why some conventional approaches to system design are inadequate when applied to complex systems.

Further, the consequences of system complexity are prevalent across a system's life cycle. Some of these consequences include high cost of development, difficulty of maintenance and repair, and the long, high cost, and the complexity of the process of system development itself [147]. Why, then, design complex systems? The reason is that an increase in complexity is accompanied by an increase in capability, at least up to a limit. As our expectations from a system grow, increased complexity results as a natural consequence. The key is to carefully manage complexity such that the value derived from marginal benefits is not overwhelmed by the costs of designing and operating such systems [157]. We revisit this in more depth in Chapter 8.

Beyond a limit, it is unreasonable to expect a single system to meet all our needs. If we set out to design a single system to fulfill our needs, we make an implicit assumption that we can find a collection of elements that, when put together, can meet our goals. In such cases, we will design the sub-systems and their interactions such that the resultant behaviors of the whole can be obtained as desired. As our needs grow, this objective becomes increasingly difficult to accomplish with individual systems with the result that we may begin to assemble independently developed systems to meet our capability needs. This brings us to the subject of a System of Systems.

1.4 BRIEF HISTORY OF SYSTEM OF SYSTEMS

Systems thinking gained traction in the twentieth century as a number of disciplines started to take a systems perspective in their problem solving. In the middle of the twentieth century, organizations started to apply systems thinking to develop more complex engineered systems, leading to a distinct discipline of systems engineering. The increasing needs of the society led to increased capability demands of new systems, and, over time, when meeting all needs from a single system became infeasible, there came to be a greater reliance on assembling existing systems, perhaps by adapting them, rather than designing new systems from scratch [145]. This development sowed the seeds of the basic ideas of modern day SoS, starting in the 1960s.

From a theoretical viewpoint, an early example of the notion of a System of Systems can be found in the ideas of Boulding, who considered it as a "gestalt" or a "spectrum of theories" greater than the sum of parts [29], and of Ackoff, who considered SoS as a "unified or integrated set" of systems concepts [7]. However, despite the early origin, the term did not catch on, and was only sporadically mentioned by other researchers through the following few decades [19]. Much later, in 1989, a mention of Systems of Systems for engineered systems can be found in the Strategic Defense Initiative [78].

The 1990s witnessed an increase in interest in studying SoS problems. During this period, two approaches to understanding SoS emerged – by definition or by characterization using distinguishing criteria. One of the earliest proposed definitions was given by Eisner, who stated that an SoS is "a set of several independently acquired systems, each under a nominal systems engineering process" [55]. More attempts at defining an SoS quickly followed. One such definition referred to SoS as a network of systems which work together to achieve a common purpose [148], while another proposed to study SoS as an artificial complex adaptive system which self-organizes on the basis of local governing rules [99].

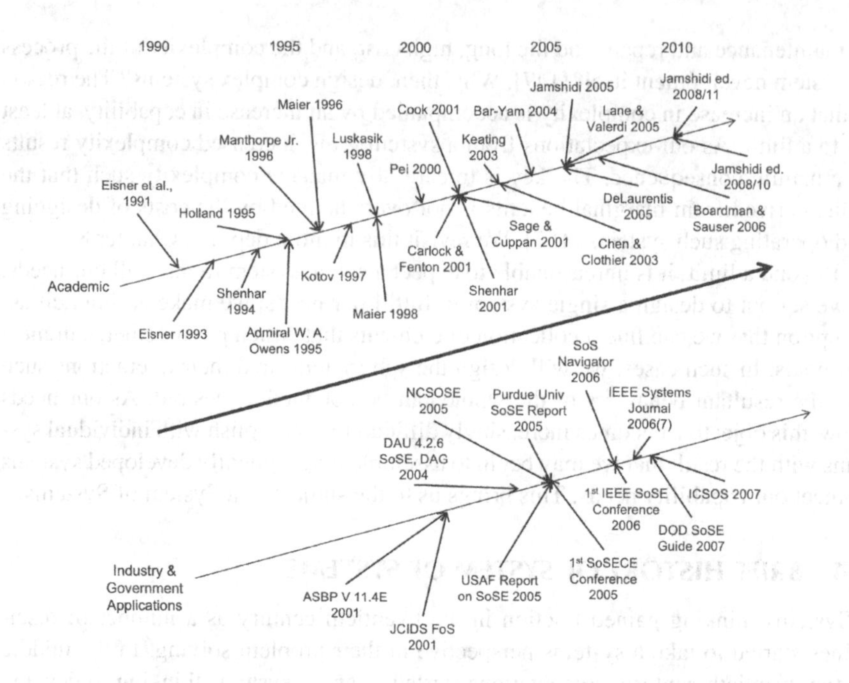

Figure 1.2: History of SoS with focus on the 2000s [76].

The 2000s brought a step-change increase in SoS, especially among industry and government, but also in academic circles and due in large part to the community building activities within IEEE of Prof. Mo Jamshidi. Gorod et al. [76] wrote an important article that reviewed this recent history, accumulating in one place many of the outputs from these different communities exploring SoS. Figure 1.2 is one of the artifacts of their review, and well demonstrates the size and richness of the activity. Yet, as we will see in the next section, defining an SoS remained a difficult task. While a single universally accepted definition of an SoS remained elusive, an alternative approach of distinguishing SoS using a set of criteria gained acceptance. Maier, for example, identified a set of five attributes, including two key ones of managerial and operational independence, to distinguish an SoS from a system. Boardman and Sauser [25] surveyed a large number of definitions to identify patterns within them which can be used to distinguish SoS; these patterns include autonomy, belonging, connectivity, diversity, and emergence.

Today, there is increasing acceptance of the need to model large systems as Systems of Systems and increasing appetite for better tools for SoS realization and continuous improvement. The government, industry, and academia have established that SoS is a distinct type of system and type of problem. Consequently, the next step is development of appropriate theory and tools for SoS-specific modeling and analysis – this book's subject– to enable the flourishing of SoSE.

1.5 WHAT IS A SYSTEM OF SYSTEMS?

A System of Systems (SoS) is a particular kind of complex system, and hence may possess many of the attributes of complex systems described earlier (see Sec. 1.3.1). The gist of what distinguishes SoS can be boiled down to two attributes of its components: they have looser, and more fluid, interconnections, and they serve twin purposes, their own and those of the SoS as a whole.

Suppose we desire a capability to surveil a portion of the coastline. To do so, we may use patrol boats, aerial vehicles or satellite imagery. Each option makes use of a different system which has its inherent advantages and disadvantages. Alternatively, since each of these systems, viz. a boat, an aerial vehicle, and a satellite, are independently operable, we can allow them to organize in a complex network whose interactions enable the desired surveillance capabilities. In such an assembly, the systems complement one another to provide greater capabilities while minimizing their shortcomings. Such an assembly is an example of an SoS.

An SoS is akin to a 'collective' – a collection of interconnected complex dynamic systems each of which is independent in structure and governance. Their need arises because the expectations of technological, social, economic, and environmental requirements in modern societies are so complex that no single system can meet them. Whereas in a system there is a hierarchy of elements with a direction of authority, an SoS is engineered such that its constituent systems exhibit synergistic interdependencies. The constituent systems maintain a certain degree of independence in their operations, and they even set their own goals which may on occasion be different from the goals of the whole. Yet, the constituent systems collaborate to achieve specific objectives which are not achievable by any of them in isolation, and, in the process, expect to gain increased value both for themselves and the collective.

While the essence of the meaning of SoS is straightforward, providing a formal definition is (and has been, per the previous section) hard. But some proposed definitions are useful and point most succinctly to the essence of an SoS. As per ISO/IEC/IEEE 21839, a System of Systems is a "set of systems or system elements that interact to provide a unique capability that none of the constituent systems can accomplish on its own." [3]. Another definition states that Systems of Systems are "large scale concurrent and distributed systems that are comprised of complex systems" [112], while yet another definition recognizes the independent purposes of constituents stating that "a System of Systems is one comprised of systems whose internal structure and exposed service set is not driven exclusively by the purpose of the larger system, or by each other; otherwise it's just a big, complex system with subsystems." [140]

This independence of systems is across all life cycle phases, from concept to operation. Each of the individual systems that make up an SoS is free to operate with its local management and seeks to maximize its own objective functions. Occasionally, this may result in constituent systems competing with one another if doing so helps them meet their own objectives. With localized management and value functions, the change in classification from a system to SoS is centered around coordination and interoperability. Independence implies that there may not be any central control authority, or, if there is one, it has very little control over the constituent systems' actions. Quite likely, if we are dealing with a collection of systems, wherein the different constituents have the ability to behave in a manner different from our commands, then we are dealing with an SoS. The amount of centralized authority available can vary and be used as a basis of classifying SoS, which we will discuss shortly.

The distinct nature of SoS also influences how they are engineered. Whereas the constituent elements of a *system* are designed and built to make a functioning whole and are not usable in isolation, the components of a *System of Systems*, on account of themselves being fully functioning systems, are built for their own purposes and they can be independently operated to fulfill certain needs. When designing an SoS, we define the inter-system interactions to get these systems to cooperate with one another, rather than design the systems themselves. The challenge during the design phase is in determining the appropriate mix of independent systems to make up an SoS, together with the network of their interactions. While selecting systems, the goal may not be to optimize system choice and architecture, rather to achieve the desired capabilities at a minimum threshold level of performance, which is called *satisficing* [149].

An additional complication for SoS arises from the fact that since individual systems are free to decide whether to participate in the SoS, the available capabilities may change or even be unavailable at times during the life of the whole. Thus, given the autonomy in choice of participation available to constituent systems, we design not just the system interactions, but also an appropriate set of incentives for the systems to continue to reliably participate as intended. This is the challenge of managing and influencing system behaviors to fulfill system-level objectives, while, at the same time, maintaining robustness and flexibility of the architecture for the SoS to be able to adapt to future evolutionary changes in participant composition. Finally, SoS invariably include humans as constituent systems, in which case the SoS is classified as a "socio-technical" system.

Figure 1.3 summarizes this discussion with perhaps the simplest depiction of an SoS that still exposes the key consequential behaviors. Collectively, the unique features of SoS contribute to the difficulties in engineering a complex System of Systems and form the motivation for Systems of Systems Engineering (SoSE); we will discuss the engineering of SoS in Chapter 2. On the topic of modeling of SoS, in Chapter 6, we will discuss *agent-based modeling*, and describe how an agent can be used to represent a wide variety of constituent systems of an SoS. In Chapter 5, we will discuss networks, which provide an effective method for representing interactions among system elements. Chapters 7 and 8 describe tools built specifically for SoS Modeling and Analysis and ways to enhance SoS engineering.

1.5.1 DISTINGUISHING SOS ON ATTRIBUTES

Our discussion so far has pointed to, but not yet offered, a useful taxonomic grouping which distinguishes Systems of Systems from other kinds of systems. Also, in Section 1.4, we described two approaches for understanding SoS, viz. by definition and by characterizing using distinguishing criteria. We will now discuss the latter

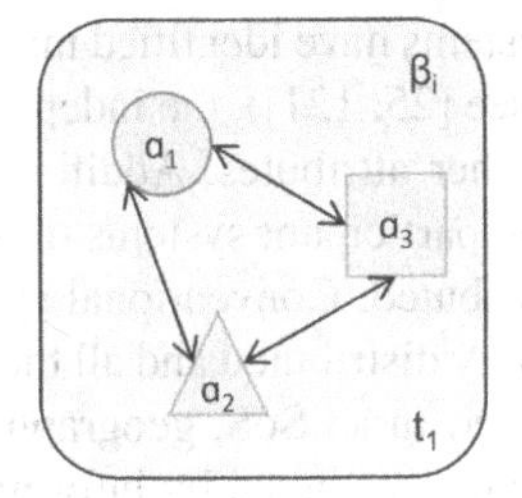

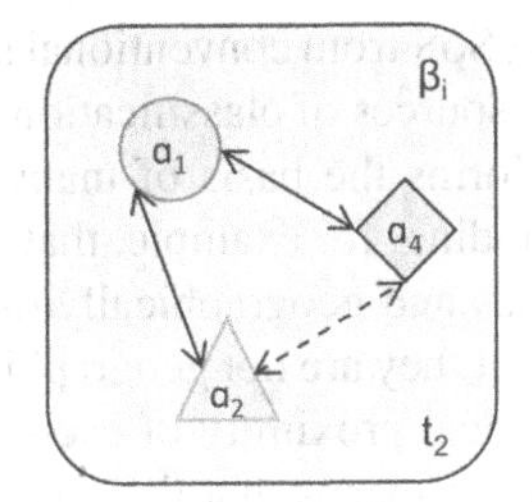

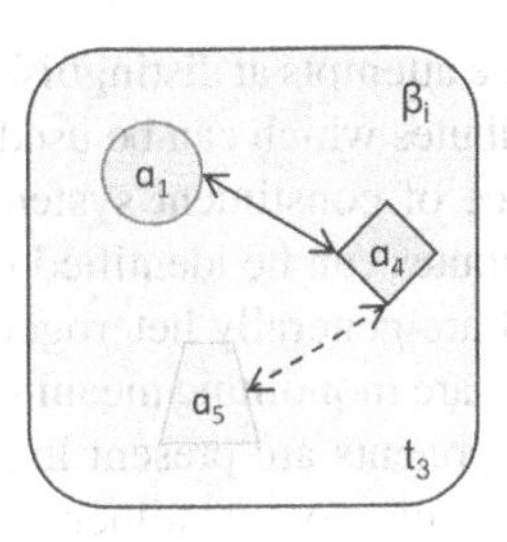

Figure 1.3: Simplest depiction of an operating SoS.

approach and look at the attributes of a system that make it a candidate for classification as a System of Systems.

The key distinguishing feature is the independence of constituent systems of an SoS – independence both in the source of control authority and in operation [121]. Given the looseness of system coupling, SoS may be regarded not as a conventional system with strictly recognized boundaries, but rather a context, or environment, of operation of multiple separate systems [6]. Each of the constituent system is a participant which remains within the boundaries of the SoS so long as it finds it beneficial to do so. Each of these participant systems is capable of independent operation even when not communicating with the other systems in the network, which means that the constituent systems continue to fulfill their functions and provide the capabilities that they were designed for, even when removed from the SoS network. Upon removal from the SoS, individual systems may exhibit reduced capabilities, but the more significant loss occurs in those capabilities that arose from inter-system interactions within the SoS environment and, in extreme cases, entire capabilities may be lost or new capabilities arise when existing systems leave or new ones join.

Similar to conventional systems, the properties of the SoS emerge from the assembly and interactions of its elements. Unlike conventional systems, however, the emergent behaviors of an SoS are not necessarily always beneficial and depend on the collective structure of interactions and composition of the participants. A key challenge to engineering SoS is to recognize emergent behaviors as early as possible and to build in only those that are desirable to SoS stakeholders. The emergent properties of an SoS may not be apparent at the time of designing or planning and would only become apparent once the systems go into operation. A cause of this difficulty is that individual systems usually have different development and operation life cycles such that the whole may not be assembled as traditional systems. Rather, constituents may be added or removed over time, sometimes due to decisions of those participants rather than a central authority. As a consequence, SoS, with their fluid boundaries, remain constantly evolving as a result of decisions of their participant systems, which is different from changes and modifications done to conventional systems, which are initiated by a single authority.

Of particular relevance to us is that modeling and simulation can prove to be useful tools in the investigation of an SoS' emergent phenomena. As we proceed to developing approaches for modeling and analysis of SoS, we will recognize that while attempts at distinguishing SoS from conventional systems have identified many attributes which can be used as sources of classification (see [25, 121]), the independence of constituent systems forms the basis of many other attributes. Additional attributes can be identified including, for example, that the participant systems of an SoS are generally heterogeneous and geographically distributed. Conventional systems are monolithic meaning that they are not geographically distributed and all their components are present in physical proximity of each other. In an SoS, geographic distribution of constituent systems means that these systems are linked by information exchange, not mass or energy exchange. As an example, the air transportation system is an SoS whose subsystems are physically separate and sometimes far removed from each other. In Part II of this book, we will present tools for modeling which are capable of capturing the key distinguishing criteria of SoS.

1.5.2 CLASSIFYING SOS ON CONTROL AUTHORITY

Independence to act in accordance with local goals implies the presence of a decision-making authority specific to a system that governs the operation of that system. The independent authority can, and often does, set its own goals and objectives and remains free to participate or leave the SoS subject to suitability to its own objectives. Such independence of decision-making, however, poses challenges to any centralized authority that may exist for control of the SoS. The scale and nature of these challenges and strategies to face them depend on the degree of control a central authority maintains. Hence, it is beneficial to classify SoS based on the degree of control available to a central authority [106, 128]. This taxonomy is also now codified in standards [5].

1.5.2.1 Directed System of Systems

The class of Systems of Systems for which the degree of control available to a central authority is the highest is called *directed Systems of Systems*. For such SoSs, while the constituent systems retain independence of operation, their objectives are subordinate to those of the SoS, and they may even be built to fulfill specific objectives of the whole rather their own. The role of the central authority is prominent both during the course of assembly and operation. In some cases even the funding for development and operation may be provided by a central authority, and the independence of management is reduced.

Figure 1.4 shows a directed SoS with multiple participating systems controlled by both local decision-makers and a central SoS-level authority. Bold, bi-directional arrows linking the central authority to the systems indicate the strong control authority of the system-level designer as compared to the weaker inter-system interactions, which are indicated by dashed arrows; inter-system interactions are largely designed by the central authority to best meet its goals. In a directed SoS, and all other following classes of SoS, the local authority has full control over its system, indicated with a bold connecting line.

An example of this type of directed SoS is an urban transportation system. The various systems and services may be independently owned and operated, but they

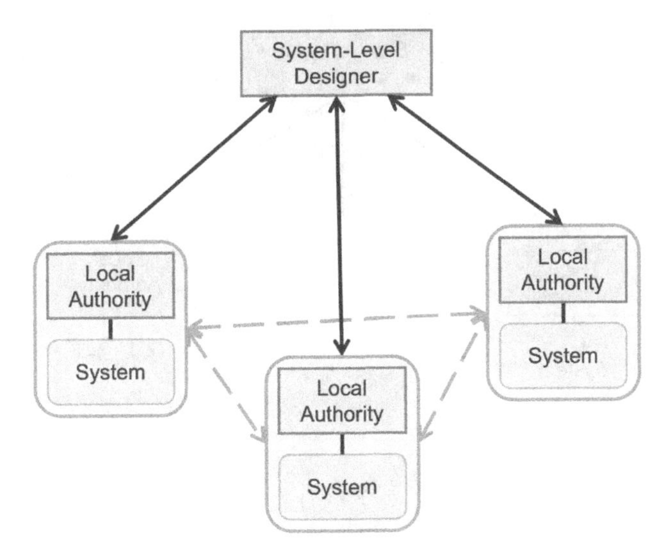

Figure 1.4: Directed SoS.

exist mainly to serve the entire transportation network. A centralized transportation authority has a high amount of control over management of operations of each of the constituent systems, for example, in setting the schedules of operation. Yet another example is a large globally dispersed logistics network. Here, whereas each of the components such as transportation systems, information technology systems, etc. may be owned independently, their operations are finely coordinated and optimized to ensure the smooth and efficient flow of goods across vast distances. In fact, a non-compliant system will likely be excluded from future operations to prevent service interruptions.

1.5.2.2 Acknowledged System of Systems

More loosely tied together than directed Systems of Systems are a category called *acknowledged Systems of Systems*. As in directed SoS, all participant systems are guided by shared SoS-level objectives, and there is a designated manager. However, constituent systems have independence of both operation and management and their objectives and funding are separate.

Figure 1.5 shows bold, uni-directional arrows from the central authority to the systems, which indicates that the central authority provides guidance to the systems, though the systems are free to make their decisions. The inter-system interactions remain relatively weak, though they are still influenced by the needs of the whole network.

The global air transportation system is an example of this type of SoS, where, for example, the participating airlines operate independently of each other and yet coordinate their actions to ensure safety and efficiency of the system. While there is

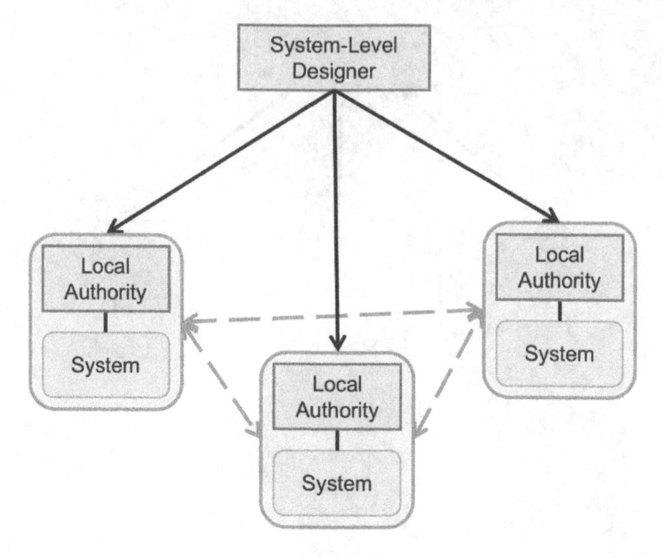

Figure 1.5: Acknowledged SoS.

no "central authority" in the sense that one body directs the actions of all participants, standards setting bodies still exist that lay down the rules which are binding on all participants. The airlines and other participants are free to maximize their utility from participation, so long as they adhere to the common rules of operation agreed by all stakeholders.

1.5.2.3 Collaborative System of Systems

Once the participation of constituent systems becomes voluntary, the System of Systems is classified as *collaborative*. There exist no central objectives, management, authority, or funding at the SoS-level, which means that each of the systems maintains a high degree of autonomy in its management and operations. The assembly is largely to address shared or common interests. Enforcement of rules and policies are based on standards rather than authority.

Figure 1.6 shows strong links between systems and their local management authority and among different systems – the bold, bi-directional arrows between systems indicates that the systems collaborate. Though there may be a central body which supports standards development, it does not act as a central authority. The structure of connections between systems is shaped by the independent decisions of each of the participating systems.

The internet is an example of a collaborative SoS, where each of the connected systems is independently operated, and they join for their own purposes, but the operation of the entire network is maintained by the use of standards.

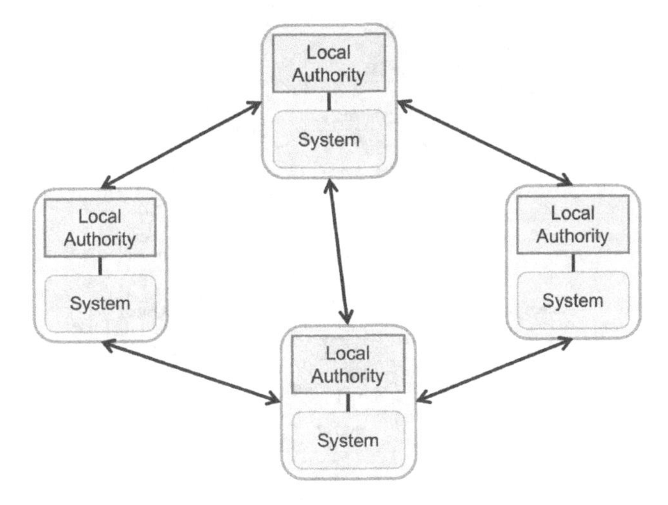

Figure 1.6: Collaborative SoS.

1.5.2.4 Virtual System of Systems

Of all categories, the *virtual Systems of Systems* are most loosely connected since there is no centrally agreed-upon purpose for their assembly. The collective behavior emerges due to decisions, including those of participating in the SoS, that each system makes for its own objectives. Hence, the inter-system interactions are indicated by dashed lines. For such a class of SoS, the observed emergent behaviors may be the most important source of identification of the boundaries of the SoS environment. No centralized control authority exists, rather compatibility of operations among systems is maintained when the systems adhere to commonly agreed standards.

Figure 1.7 shows the SoS as a collection of systems with the weakest interactions of all classes of SoS, indicated by dashed links. No central authority exists in this case. Instead, a cloud in the background of the systems indicates that common standards guide system behaviors and interactions and that the boundaries of such SoS are the most ill-defined.

While we classified the internet, which is a network of connected devices as a collaborative SoS, the World Wide Web (WWW) is an example of a virtual SoS. The WWW is a network of connected documents and web-pages and there is no single central authority which guides its architecture or behavior. Rather, all websites which constitute the WWW comply with common interface and data standards, thus enabling a connected whole to function.

1.6 CHAPTER SUMMARY

In this chapter we presented some definitions that will now be used throughout the rest of this book; these definitions are summarized in Table 1.1. The key ideas presented in this chapter were:

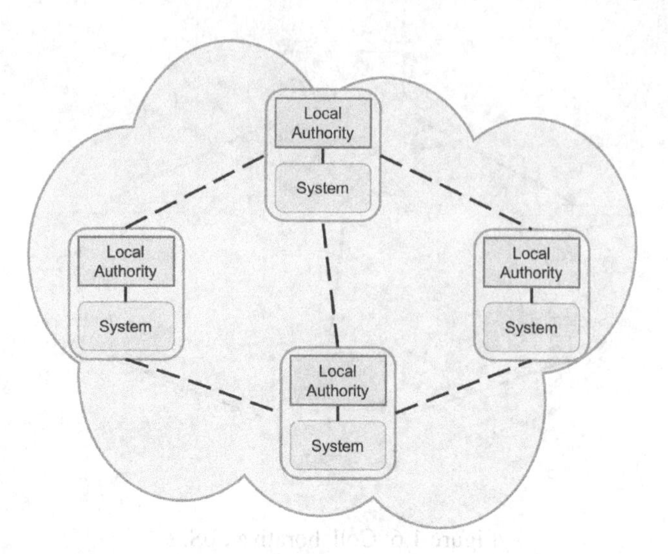

Figure 1.7: Virtual SoS.

1. The reductionist approach to problem solving is to reduce the size of a problem to decrease its complexity. In contrast, a holistic approach takes a "systems thinking" approach and is useful for the design and development of large-scale complex systems.
2. Systems have the properties of a collection of elements, interdependence among its elements, and a goal for which they were designed. Complex systems additionally display emergent behaviors which are difficult to model and predict.
3. Systems of Systems are a type of large-scale complex systems with looser, more fluid interconnections among the constituent elements. The constituent elements are themselves useful systems in their own right and maintain a degree of independence higher than that found within the elements of a system.
4. The constituent systems within an SoS have both operational and managerial independence. Operational independence means that the systems have their own useful purposes outside the SoS, while managerial independence means that the systems are provided their own unique purposes by owners and operators.
5. A systems approach is to identify a system and explain behaviors of properties of the whole and the components.
6. Systems engineering is a structured approach employed for the design and development of systems; it is an application of systems thinking for the purpose of development of a system.

Table 1.1
Summary of Systems Definitions

Term	Definition
System	A set of different elements connected or related so as to perform a unique function not performable by the elements alone.
System of Systems (SoS)	Systems of Systems arise when constituent elements are themselves useful systems in their own right and maintain a degree of operational and managerial independence in addition to participation in the larger collective.
Systems engineering (SE)	Systems Engineering is an interdisciplinary approach and means to enable the realization of successful systems.
Modeling	Modeling is the process of developing models of a system, whether physical, mathematical, or computational.
Analyzing	Analyzing is the process of simulating a model and observing its response to the given inputs.

1.7 DISCUSSION QUESTIONS AND EXERCISES

1.7.1 DISCUSSION QUESTIONS

1. Which of the following are ways to describe the contents of a wooden chair?
    a. Lignin and carbohydrates
    b. Atoms of (primarily) carbon, hydrogen, and oxygen
    c. Fermions and bosons
2. It is appropriate to call an assemblage of interconnected components a system if:
    a. It has a goal or purpose (as an assemblage)
    b. Components have operational independence
    c. The assemblage evolves over time
    d. None of the above

1.7.2 EXERCISES

1. (This is an individual assignment.) Review the "Solberg Chart", Figure 1.1, and any other SoS-relevant "system theory/method" that does not appear explicitly on it, e.g., Reinforcement Learning, Block Chain, Mechanism Design, and Consensus Control. See Chapter 4 for brief information about all topics listed on the Solberg Chart. Select one of the methods (or a method) and answer the following questions:

    Question 1: Who created the method and in what year?

Question 2: What are the key mathematical principles used in the method?

Question 3: What are well-known applications of the method?

Question 4: What are weaknesses/limitations of the method?

Question 5: How does this approach apply to generalized System of Systems Engineering?

2 What Is System of Systems Engineering?

2.1 OVERVIEW OF *SYSTEMS ENGINEERING*

We defined a system as a collection of elements which provides capabilities that no single element can provide alone. The additional capabilities, beyond that of individual elements, are the result of interactions among those elements. We intentionally exploit this quality of systems, that of providing greater capabilities than those that their elements provide in isolation, by bringing together a set of elements and designing their interactions to achieve desired capabilities. However, as systems increase in complexity, we need to assemble larger numbers of elements, account for more complex and interaction patterns and policies, and try to drive the system towards behaviors which are generally non-linear and become increasingly more difficult to predict. The increase in system complexity, therefore, also increases the value of using principles and patterns of systems thinking for design and engineering of systems. Application of systems thinking provides us with a structured approach for designing and building better systems to fulfill our needs; this is referred to as Systems Engineering (SE):

1. "[Systems Engineering is] an interdisciplinary approach and means to enable the realization of successful systems." [167]
2. "Systems engineering is a process employed in the evolution of systems from the point of when a need is identified through production and/or construction and ultimate deployment of that system for consumer use."[22]
3. "Systems engineering is a robust approach to the design, creation, and operation of systems." [97]

An SE approach helps satisfy the needs of not just the end users, but those of all stakeholders including the organization building those systems, the operators, and even the regulators. Frequently, different stakeholders have differing expectations of the system. For example, from the customer's viewpoint, the purpose of SE is to find the best solution to their problem, while from the organization's viewpoint the purpose is to manage the risk involved during the process of building systems [33]. Meeting different stakeholders' needs involves ensuring that the organizational, cost, and technical aspects are balanced against each other, even when such needs are conflicting with one another [97]. When successfully executed, SE enables us to derive the highest value for all stakeholders involved.

Every system development endeavor begins with the identification of a need which could arise from a desire for a new capability, development of new technology, or even a need to replace a legacy system. A formal systems engineering process begins with identification and transformation of needs into technical requirements. The beginning of a system development program involves tasks such as market study, feasibility studies, etc., with the objective to clearly define the problem. From the customer's needs we derive a set of technical objectives which specify the specific target levels of outputs that the system must achieve [97]. We transform these objectives into a set of validated technical requirements which are statements which describe system inputs, outputs, and functions that a system must fulfill in order to meet customer needs. Only after the needs are rigorously evaluated, requirements clearly defined, and solution feasibility established, is the program initiated.

Thereafter, the process advances through the selection of criteria for evaluation of candidate solution alternatives, synthesis and analysis of alternative solution concepts, design iteration and optimization, selection of a final concept, and manufacturing and deployment. Early in the development phase, engineers will propose multiple alternative solutions. The proposed alternatives may use a combination of new technologies or reuse of existing systems in different architectural layouts to arrive at designs which meet requirements. This process is called "system synthesis" and it leads to a set of feasible alternatives which should satisfy the design requirements. This is accompanied by the process of system analysis where we evaluate how well each of the proposed alternatives performs relative to objectives. Analysis is used to compare designs, identify system behaviors, and guide decision-making. This requires the selection of appropriate measures of performance against which the system will be measured.

In the above brief description of the SE process, we have described a sequence where, as we advance, the design detail and our knowledge of system behaviors become more refined until we arrive at a final configuration. But this process is rarely, if ever, linear. The entire process proceeds in a sequence of iterating loops. The iterations involved are integral to this approach, because system design is a process of discovery of the solution. For all systems, but especially for larger systems, the iterations can happen from the beginning of the development to their operation. During refinement, we optimize system design, change configuration based on newly acquired knowledge, or even update requirements to better suit the problem definition.

Figure 2.1 shows a representation of the systems engineering process, called the Vee-model. The above steps form the left arm of the Vee-model of SE; that is, the tasks of specification of requirements, synthesis of alternative solution concepts, their analysis and further refinement are successively applied at each level of system hierarchy starting from the whole system down to its lowest subsystem element. The right arm of this model shows the process of system realization, which involves the steps of integrating subsystem elements into higher level subsystems leading to the complete system.

Connecting the two arms of the Vee-model are the processes of verification and validation. Verification answers the question, "Did we build the system right?", and

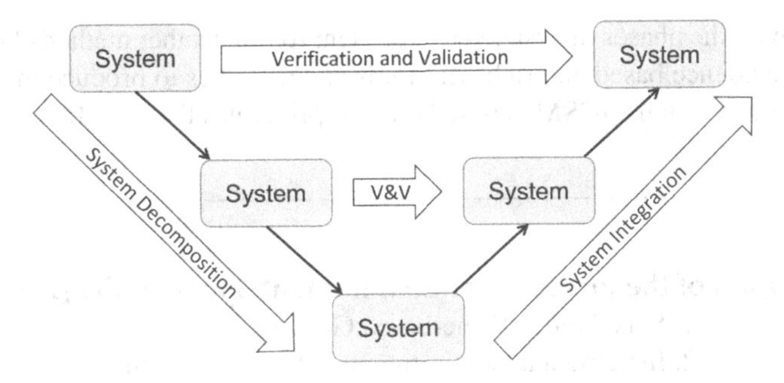

Figure 2.1: The Vee-diagram of SE, a standard representation of the SE process.

checks to see if the set of system elements and their planned interactions were properly implemented. Validation answers the question, "Did we build the right system?", and takes the customer's viewpoint and checks to see if an appropriate solution to a need was delivered. The Vee-model is one of many alternative representations of the SE process; others include the spiral model, waterfall model, etc.

While most attention is paid to the process of development until release to manufacture, systems engineering is applied until the end of a system's life cycle. In fact, for large systems with long life cycles which can stretch over decades, changes to design may be done even while the system is in operation. A typical example of this is an aircraft for which it is not uncommon to undergo mid-life upgrades to extend its operational life. Major upgrades may be significant engineering undertakings which may keep the aircraft grounded for a long period of time and will require its own systems engineering program. Depending on the scope and complexity of an SE program, the entire program life cycle can be divided into a series of phases which begin from identification of need and end with system disposal. One way to break down a program into a sequence of phases is shown in Fig. 2.2, though other ways of specifying phases exist ([97, 167]).

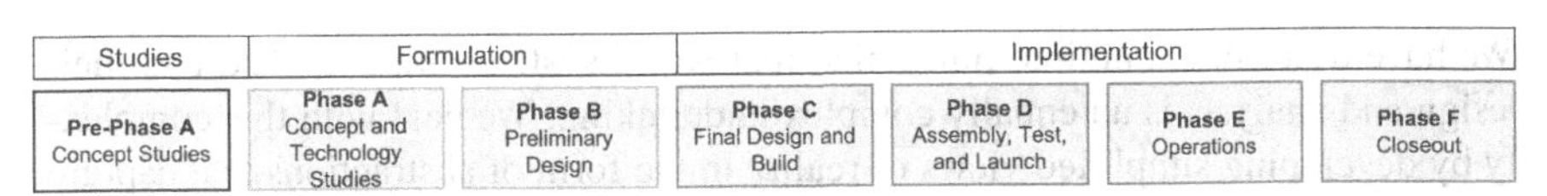

Figure 2.2: Phases in a Systems Engineering process.

The more complex the system being engineered, the more difficult will be these decisions along the life cycle. In addition, we grant that many of the examples, and much of the context, in this text have a "hardware flavor", perhaps unavoidable given the authors' backgrounds. However, software is as critical, if not more, in many emerging systems. The amazing increase in the numbers of lines of code involved in the 3rd, 4th, and 5th generation fighters illustrates this statement. In these cases especially, there has been an important development in recent years in the form of the Incremental Commitment Spiral Model (ICSM) of Boehm et al. [26]. ICSM is congruent with the phases of system development found in other methods but, importantly, it is evidence based and rich with tools and activities to produce the evidence for risk-based decisions. ICSM is based on four principles listed in Table 2.1.

Table 2.1
Four Principles of the Incremental Commitment Spiral Model [26]

1. Stakeholder Value-based Guidance
2. Incremental Commitment and Accountability
3. Concurrent Multi-discipline Engineering
4. Evidence and Risk-Based Decisions

The basis in evidence and the four ICSM Principles together have strong correspondence to the principles embedded within the SoS Modeling and Analysis process to be presented in Chapter 3 and the overall approach to SoS problems in this text. Since we do not emphasize software development in SoS in this text, we do suggest to the reader a deep exploration of the ICSM to attain specific knowledge and skill for software-intensive SoS design and architecting decisions.

Finally, we note that, in addition to the technical, SE also includes tasks related to project and personnel management. The non-technical aspects of SE are as important as its technical aspects and, together, serve as a logical basis for system development and operation. In all its phases and for all tasks where it is applied, SE aids by systematically capturing new knowledge at every step of the process and guiding future actions towards system development. It provides an ability to tackle complexity, verify assumptions, gain a better understanding of client needs, manage change and configuration flexibly, and produce the highest quality, economical, and efficient solution possible.

2.2 ROLE OF MODELING AND ANALYSIS

We have described, multiple times, that real-world systems are complex, and their design and analysis is an equally complex undertaking. We deal with this complexity by developing simplified views of reality in the form of abstractions that capture only those features of the system's structure and behavior which are relevant to our task and leave out the rest. Everything outside the boundaries of the system is either neglected or only included when analyzing the system's interactions with its environment. Such an idealization of a part of the real world is called a *model*.

Modeling is the process of developing abstractions of reality to represent only the portion that we intend to reason about. The decision of what to include and what not to include depends on a judgment of the degree to which a certain behavior will affect our analysis. For example, when modeling the motion of a train, we can leave out the effect of aerodynamic drag and still calculate a fairly good value of its speed given the forces applied by its engine. While doing so for an aircraft, leaving out drag will give an inaccurate result.

The form of model we develop will depend on its intended use and available resources. Choices available include representing a system with mathematical equations, in the form of a physical artifact, or even as a computational model. The aim is to develop a representation that can be used to evaluate the performance of the system under development. Since a model is meant to be an abstraction of a system and not its complete representation, we must select the form of the model which allows us to reason about the available alternatives. Generally, this means developing custom models for the problem at hand to capture the essential elements of a system and the behaviors that contribute to the system's value. A good model has various qualities:

1. It will represent the system, including the constituent elements and their interactions that are essential to explaining the system's behaviors. Further, the model will capture the rules and policies that guide the constituents' behavior and operation.
2. It will be at the right level of abstraction, which means that all essential details are captured while none of the extra information is included. That is, a good model will focus only on those aspects of the system which are relevant to the problem at hand.
3. It will be at the right level of fidelity, which means that it will be detailed enough for us to derive insights, differentiate alternatives, or make predictions using it, and yet simplified enough for us to both manage and use it. Even with a given level of abstraction, fidelity can differ based on, for example, the choice of mathematical equation we use to represent the system.
4. It will be flexible and can evolve as more information becomes known.
5. It will provide support for problem analysis and decision-making [59]. This includes providing guidance on the use of development processes, the set of objective functions, the type of system to be developed, or even how much of the development is new versus how much is reused from previous systems must be considered.
6. It will be represent different organization structures and show how product architecture and organizational structure influence one another [97].

Bearing in mind the above qualities of a good model, there are certain good practices to keep in mind when developing models:

1. Build models with minimal complexity to reproduce key behaviors. That is, identify a simple model with minimal set of parameters that demonstrates the behaviors that we seek and identify elements of those models which are necessary and sufficient to produce that behavior. This is a holistic approach, i.e., looking for similarities between systems rather than between their parts.
2. Test thoroughly and revise the model as required. This requires understanding logic of system architecture and its behaviors.
3. Add complexity as necessary. As the knowledge of a system increases, the fidelity of representation can be increased. A highly detailed model of a

system may allow us to reason accurately about its behaviors but likely at a high expense. When developing models of the whole system, therefore, we may choose to sacrifice realism of system representation in favor of simplicity of analysis.

Ultimately, we intend to use these models to either explain the current states of reality or predict future outcomes. Following the development of a system representation, this is done via the steps of model execution and analysis [126]. Model execution means using them for design or analysis of the system, capturing the interrelationships among its components, or even communicating with a system's stakeholders. Depending on the type of model, execution can be in the form of deductive reasoning as in the case of mathematical models, experimentation using physical prototypes, or simulation of computer models.

Analysis is the process of simulating a model and observing its response to the given inputs. As per ISO/IEC/IEEE 15288, "the purpose of the System Analysis process is to provide a rigorous basis of data and information for technical understanding to aid decision-making across the life cycle" [2]. Analysis supports the process of system synthesis in decision-making, identification and evaluation of alternative solution concepts, or in understanding the behaviors of a particular system architecture. The process of analysis happens throughout the life cycle, alongside synthesis in iterative design loops. We use the developed system models for analysis of proposed designs, which helps reduce risk and program cost.

Modeling and analysis takes on an added importance for large complex systems such as SoS. While traditional fields of sciences relied on use of scientific laws that could be expressed as mathematical equations which could be solved analytically or numerically to derive insights, for larger systems such as SoS, such laws are frequently absent. Hence, for such systems, we simulate rather than solve models at a suitable level of abstraction. The insights derived are as much or more a function of the analyst as it is of the fundamental attributes of the system itself. Because SoS are inherently complex, development of deterministic models is difficult, if not impossible [14]. We will discuss modeling approaches applicable to SoS in Section 3.1.

But what about *optimization?* Certainly, it is an expansive question. In fact, optimization is one of the most popular means of concluding the analysis process we just discussed, in particular focused on bringing about the best design alternative. Indeed, if we claim that SoS M&A has the goal of making better decisions (especially for the long term), how could the concept of optimization *not* play a central role? Over the past two decades, as the concept of SoS has emerged into wider circles, this question about the role of optimization has been vigorously explored, at times even argued! It is worthwhile to consider a brief snapshot of some of this sub-history before we move on to the wider topic of system engineering, especially the associated modeling and analysis, for SoS.

Among the numerous distinguishing features of SoS, the hierarchical nature consisting of distinct types of component systems prompted the Multidisciplinary Design Optimization (MDO) community to engage in the question of optimization and SoS. Nowadays, more often called Multidisciplinary Design Analysis and

Optimization (MDAO), this community was an early home for DeLaurentis and indeed contributed positive momentum to the SoS world. One of the earliest works in this genre was Dennis et al. in 2005 [50]. The importance of this work, authored by members of the well-known MDAO research group at the Boeing Company, lay with the clarity on how one would use a canonical MDAO framework and apply it to an SoS design problem, tailored by the degree of "activity" of what they termed the Central Authority in the SoS. Later work by Jaroslaw Sobieski, one of the founders of MDAO, sought to leverage the hierarchical structure of MDO problems, where the system level sought to both integrate and converge with a lower level of distinct component systems, and apply it to SoS [152]. A variety of other accomplished MDAO researchers have also contributed in important ways, including especially Profs. Papalambros and Kokkolaras, for example. Their work successfully translated the concept of target cascading (and uncertainty in the targets) into SoS problems with definitive hierarchy [110, 125].

A limiting factor, however, in adopting these MDAO approaches to SoS lies with the essential distinguishing traits of SOS: operational and managerial independence. In almost all MDAO applications (of awareness by the authors), the contributing disciplines are not independent agents; they are some form of mechanics or physics. For example, in the classical MDAO application of aircraft wing design optimization, the contributing disciplines are aerodynamics, structures, and control. These three do not need to be "incentivized" to participate in the SoS, nor does the degree of intelligence in their actions need assessment. They have no independent behavior other than the physical laws that define them. For sure this is a simplified view, yet we believe still an accurate perspective that is at least a limiting factor in use for SoS M&A. In addition, as we have already seen in Chapter 1, the very nature of SoS, especially those instances in which there is very little control at the SoS level, makes the goal of optimization less powerful than that of sacrificing.

2.3 SYSTEM OF SYSTEMS ENGINEERING

2.3.1 WHAT IS DIFFERENT?

Systems engineering is typically concerned with development of a monolithic system (product or process) with well-defined boundaries. These problems are still hard – after all, many are complex systems where defining requirements and developing robust solutions is challenging. Examples include the development of a commercial transport aircraft, a tower/large building, or a rocket as in Figure 2.3(a). In comparison, Systems of Systems Engineering (SoSE) is the process of assembling a mix of existing and new systems into a new whole with capability that is greater than the capabilities of its constituent parts. It strives for effectiveness in development of architecture for a network of independently operating systems that can collaborate and collectively provide unique capabilities that arise from their interactions. An air transportation network, a constellation of space exploration vehicles, or a smart city are examples that require SoSE due to these additional considerations in their design and development as in Figure 2.3(b).

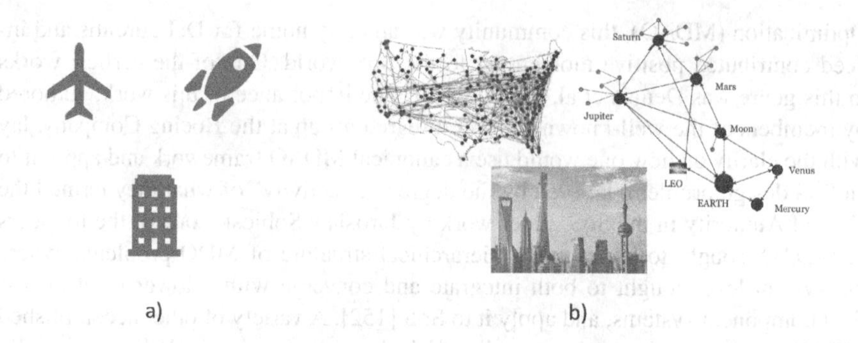

Figure 2.3: Examples that illustrate the difference between (a) SE for monolithic systems, (b) SoSE for integrated, collaborating, and independent systems.

SE is based on the assumption that given a set of requirements, engineers can build a system to meet those requirements. This means that given a set of technical requirements, designers can select a set of components which can be assembled into an integrated whole such that they will work together to provide the desired capabilities. However, this assumption breaks down if the boundaries of the system are open or ill-defined, the behavior of components unpredictable, and requirements not clearly defined or unknown. Large uncertainties in nature of component interdependence or in ways to predict the evolution of these interactions further complicates the task of system development. While some authors have found means to use the SE Vee (see Figure 2.1) for SoSE, such as [98], the specific modeling and analysis underpinning SE decisions are challenged by SoS even when the steps involved are the same.

Besides providing a clear taxonomic node for distinguishing SoS, Mark Maier's seminal paper [121] also provided four design heuristics that remain, in our minds, as essential for shaping SoS M&A in support of design and 'engineering' in the broadest sense. In Table 2.2, we list these four heuristics and relate them to themes, approaches, and insights related to M&A that will be treated in later chapters.

Let us close this subsection with a reminder of something quite basic: We are interested in developing and operating a System of Systems expressly because we want capabilities not obtainable from individual systems. As focus shifts from development of individual systems to integration of multiple independently developed systems, new challenges relating to management of relationships and interactions between these systems emerge across the life cycle. "Non-SoS-tailored" systems modeling, analysis and engineering will likely hit limits of managing increasing mission complexity. Framing of the design problem and representation of the system no longer remain as clear or precise for SoS problems [34, 59]. In this context, the systems generate their elements, and not the other way round. We need a new way to engineer systems when the requirements are not easily decomposable, i.e., a way of thinking about systems development, tailored for SoS.

Table 2.2
Relating Maier's Four Heuristics for SoS Architecting and their Implications for SoS M&A

Maier Heuristic	Implication for SoS Modeling and Analysis
Stable Intermediate Forms (SIFs): complex systems evolve faster with SIFs; establish early (valued) capabilities, then evolve.	Need to understand what SoS participants value, and how to recognize and evaluate a stable capability.
Policy Triage: choose carefully what you try to control; standards; be willing to cut out or add.	Policies and levels of autonomy must be design variables and thus reflected in SoS models.
Leveraging at the Interfaces: the greatest leverage in system architecting (and greatest danger) is at the interfaces.	Models of intended component interactions (and ability to detect unmodeled ones) are critical to manage emergent behavior.
Ensuring Cooperation: since systems can choose to participate (or not), incentives must be intentionally designed.	Modeling is the process of developing models of a system, whether physical, mathematical, or computational.

2.3.2 SYSTEMS THINKING: A KEY TO SOSE

Once we have recognized SoS a distinct sub-class of complex systems, we need to extend systems thinking to account for the additional features exhibited by them. What we need is a comprehensive approach to model SoS, comprehend the design space, and synthesize practical SoS solutions. In the case of the SoS problem, it is probably the case that there is first and foremost a need for a broader methodology to help decide which tools – from operations research, economics, engineering, etc. – are needed to perform a similar dimensional collapse of the problem space for the particular problem being considered.

Surely there are by now a myriad of processes for SoSE that various organizations may have, perhaps many referred to by a different name and less formal than other SE-related processes. However, while the goals of any design process tailored for SoS would be similar to those of traditional system development, that is, delivering the desired capabilities via assemblies of systems at high performance levels, the application of a systems approach to Systems of Systems involves identification of the following features of SoS and the problem of their design and operation:

1. They are assemblies of large-scale independent systems which show interdependent behaviors. During design as well as operation, no central authority has absolute control over all components of the SoS. Rather, it provides guidance in the form of influence over the component systems.

2. The range of timescales of design, planning, and operations is much bigger. Collaboration among participating stakeholders needs to support both new entrants as well as legacy operators, and the evolution of the network due to addition, removal or change in participating systems.
3. Designers have greatest leverage over the interfaces and very low, if not none, direct authority over the independent systems. Identification and measurement of interfaces and emergent behaviors are the most important contributions that a centralized designer can do for the SoS.
4. The 'variables' of design are ill-defined, constantly changing, and incomplete. Uncertainty exists across the entire process – from the knowledge of requirements to the list of objectives.
5. There are aspects of design which cannot be quantified. Consideration of social phenomena in addition to technical ones is crucial.

System independence, which exists throughout the life cycle from design to operation, requires a new approach to designing new systems with higher autonomy given to organizations developing each system. Traditional systems engineering does not provide us with guidance on development of SoS since it presupposes a central authority with absolute control over all aspects of the design process [106]. In case of SoS, independent stakeholders contribute to the development and operation of an SoS. Every participating organization has its own policies and processes. Systems thinking applied to Systems of Systems needs to forgo any reliance on "total control" of the process [145]. At the same time, any central authority needs to move away from a traditional view of the system as a single whole, to one where the constituent elements maintain a high degree of local power, even after they begin operations.

Independence of the stakeholders further means that all information is compartmentalized and we need appropriate incentives to encourage participation especially if the stakeholders have competing interests. Thus, a systems approach for SoS will need to address issues such as who controls the data, or how we resolve competing issues among stakeholders. When setting up incentives for SoS, we must acknowledge the local goals and agendas of participating systems.

Modeling and simulation capabilities for SoS need new information/knowledge integration techniques to assist in the SoS management, operation, and control. Uniform methods to information and knowledge management will help all stakeholders take a consistent view of the whole and develop whole system analysis techniques. Advancement in modeling and simulation techniques will also help lessen the difficulty of testing and validation including a lack of approaches for identifying emergence. Due to lack of known principles of SoSE, which stems at least partially from a lack of theoretical grounding, systems thinking for SoS can also support the development of established systems approaches such as those for global optimization to support optimization of SoS management, operations, and control through optimal policies/rules/governance. Even extending the systems lexicon and taxonomy will be beneficial, for example, in modeling of human and social aspects of an SoS which is constrained by a lack of common vocabulary and data [137].

The current SoSE methodologies, where systems approaches have been applied to SoS, largely provide high-level processes and guidance that are aimed at orchestrating activities for a particular domain (defense, transportation, etc.). The SoSE processes that aim to be domain-agnostic and emphasize defining knowledge artifacts and contexts that may not be easily constructed and replicated are still lacking. Current methods provide no support to address pertinent questions such as what are knowledge artifacts that are needed when the intent of the SoS evolution is to innovate, what artifacts and processes are needed to address the high degree of uncertainties associated with such types of evolutionary intent that projects far into the future, and how can SoS solutions be effective given that end solutions may be adopted only on a voluntary basis at the local level?

2.3.3 DISTINCTIVE FEATURES OF SOSE

Now that we have discussed the extension of systems thinking to SoS problems, let us look at how Systems of Systems Engineering will differ from traditional systems engineering. In brief, some differences between SoSE and SE include a lack of control over individual systems versus complete control at all stages of design and operation, local objectives of participating systems versus a hierarchical ordering of requirements, asynchronous and variable development times versus coordinated design, assembly, and operation.

As we stated earlier, in SoSE, a central authority does not assume complete control of participating systems. In general, each of the participating systems may be designed to meet its respective needs with little expectation of being subject to the needs of another, higher level authority. The individual stakeholders may have interests beyond SoS, and the engineering approach must balance the two. Further, each individual system may be heterogeneous and can be partitioned into various multilayered networks. An SoS engineer may have no say in the structure of those systems except with regards to configuration of their interfaces.

Not only would the structure of each system differ, the systems are also likely to have different development and operation timescales. SoSE needs to account for the planning of individual systems including the possibility of asynchronous updates and changes to participating systems. At the start of a development program, SoS may be assembled from a combination of systems which may or may not exist yet, or may be partially developed with uncertain life cycle time. Thus, the engineer must also anticipate changes in the set of participating systems due to addition of new systems, removal of old ones, or changes to those within the network.

Finally, whereas traditional SE aims to deliver a system which meets customer needs, the outputs of an SoS engineering program may not be tangible systems at all [59]. Instead, the outputs may be the results of analysis of system-level behaviors or an understanding of emergent properties. Models supporting SoSE must allow for and help humans understand emergent behavior (either good or bad), especially in applications where failure carries very high consequences in cost or capability. The ability to predict effects of autonomous decision-making of systems on the rest of the SoS will be especially valuable.

To summarize, some of the constructs that we must understand deeply to model SoS include:

- System capabilities: behaviors
- Interactions: static, dynamic, directed, weighted
- Interfaces
- Policies: forms of interactions
- Boundaries
- Layers: aggregation, abstraction

2.3.4 SOSE: AN INDUSTRIAL SNAPSHOT

While this book offers general principles, methods and processes for SoS M&A, which is the core activity within the larger SoSE activity, we recognize that there has been important progress on adoption of SoSE by industry. The evidence lies in three streams: standards development, industrial practices, and application-oriented R&D / testing. Much of this is out-of-scope for our treatment. However, it is important nonetheless and thus worthy of some highlights and pointers.

First, a small number of standards have been published for SoSE. According to SEBoK [24], as of 2022 three standards for SoSE have been adopted by the International Standards organization. These came about primarily from the report of an ISO SoS Standards study group (ISO, 2016) recognizing the increased attention to SoS and the value to standards to the maturation of SoSE. Following are short capsules on the three relevant standards as summarized in SEBoK:

ISO/IEC/IEEE 21839 – System of Systems (SoS) Considerations in Life Cycle Stages of a System [3] – This standard provides a set of critical considerations to be addressed at key points in the life cycle of systems created by humans and refers to a constituent system that will interact in a system of systems as the system of interest (SOI). These considerations are aligned with ISO/IEC/IEEE 15288 and the ISO/IEC/IEEE 24748 framework for system life cycle stages and associated terminology.

ISO/IEC/IEEE 21840 – Guidelines for the utilization of ISO/IEC/IEEE 15288 in the context of System of Systems (SoS) Engineering [4] – This standard provides guidance for the utilization of ISO/IEC/IEEE 15288 in the context of SoS. While ISO/IEC/IEEE 15288 applies to systems (including constituent systems), this document provides guidance on application of these processes to SoS. However, ISO/IEC/IEEE 21840 is not a self-contained SoS replacement for ISO/IEC/IEEE 15288. This document is intended to be used in conjunction with ISO/IEC/IEEE 15288, ISO/IEC/IEEE 21839 and ISO/IEC/IEEE 21841 and is not intended to be used without them.

ISO/IEC/IEEE 21841 – Taxonomy of Systems of Systems [5] – The purpose of this standard is to define normalized taxonomies for systems of systems (SoS) to facilitate communications among stakeholders. It also briefly explains what a taxonomy is and how it applies to the SoS to aid in understanding and communication.

Second, the increase in formal development and application of SoSE methods, processes, and tools by the industry (especially in the aerospace and defense domain) indicates the maturation of the practice. One can easily find small, medium, and large companies that tout their capabilities and successes in applying SoSE. The regular appearance of experts from these organizations at technical conferences and their consistent desire to hire graduates with SoS M&A skills also lends credence to the vibrancy of SoSE in practice.

Third, a fair number of examples exist of applied research and development especially as it relates to the low level interoperability challenges in SoSE. In particular, the U.S. Defense Advanced Research Projects Agency (DARPA) has executed a number of inter-related programs over the past decade in pursuit of concrete demonstrations of SoS-style interoperability, meaning an increasing capability for systems to interact in multiple modalities and in a dynamic, adaptive, or even ad-hoc manner. One can find descriptions of these programs at DARPA's website (https://www.darpa.mil/) or in the available literature on results from these efforts. The Federally Funded Research and Development Centers (FFRDCs) and DoD-sponsored University-Affiliated Research Centers (UARCs) communities have also made significant contributions to advancing both the development and practice of SoSE. Very prominent among these is the work of Dr. Judith Dahmann and colleagues at the MITRE Corp. While we only reference a small number of her works, almost all of them collaborative, we could have included many others. Further, many of her works appeared at the important early phases of the recent emergence of SoS and helped the community come to terms with research needed, and the language for articulating that need (see especially [40]). As important, Dr. Dahmann has dedicated much time attending academic and professional society conferences and thereby providing indispensable conversations, insight exchange, and enrichment to the combined efforts of the SoS community.

2.3.5 SOSE IN ACADEMIA: SMALL START, STEADY GROWTH

Some of the early conceptions of SoS covered in Chapter 1 (and indicated in the upper portion of Figure 1.2) came from academic researchers. However, it wasn't until the late 1990s and early 2000s that SoSE-related topics saw tangible roots in both academic research circles and curriculum. One of those places was the University of Southern California (USC) Systems Architecture group pioneered by Prof. Eberhard Rechtin and colleagues. They began introducing specific content on SoS into their master's programs in System Architecting in the late 1980s. A nice paper was published in 2005 documenting the findings and implications of this introduction, summarizing the excitement and benefit that students self-reported related to the new SoS content [38].

Courses and research programs at numerous universities presently exist that, while not all entirely dedicated to SoS (as the Purdue AAE 560 course is), nonetheless have significant and explicit SoS context. For example, CE397 – Urban Systems Engineering taught by Prof. Kasey Faust at The University of Texas at Austin introduces SoS problem formulation and modeling methods as key approaches in achieving high performing and sustainable urban systems. Another example is the research program and courses taught by Prof. (Datu) Buyung Agusdinata taking SoS lens and approach to examine a wide variety of challenging domains including system modeling food-energy-water nexus, sustainable transportation including alternative fuels, and several others. Finally, the work of Prof. Ali Raz at George Mason University is advancing the integration of Artificial Intelligence (AI) into SoS and information fusion applications. Of course, George Mason University is notable for the historic contributions to Systems Engineering (notably the late Andrew Sage) and Architecture (notably the contributions of Alexander Levis, his academic output and his shaping of the early SoS innovation landscape when serving as Chief Scientist of the U.S. Air Force from 2001 to 2004).

2.4 EXAMPLE: AIR TRANSPORTATION SYSTEM AS A SYSTEM OF SYSTEMS

Consider a national air transportation system, for example that of the United States. Many different airlines operate numerous aircraft on an ever-changing set of routes across the country. For their operations, the airlines make use of airports along with their associated ground facilities. The airlines are independent in their operations, meaning that they make their decisions to fulfill their own objectives regardless of those of the other airlines, yet all must comply with certain system-wide constraints. However, it is only by working together, all the airlines, aircraft and ground-based infrastructure, that the capability to transport passengers across the length and breadth of the country is achieved. Thus, the Air Transportation System (ATS) is a collection of systems which are operationally and managerially independent, and can therefore be classified as an SoS, in particular an Acknowledged or even Collaborative one.

The capability achieved by this SoS is staggering. In 2019, the total number of passengers carried in the United States alone was 860 million. There are over 5000 public-use airports, heliports, and seaplane bases in operation in the US, and many more for private use. The US airlines collectively generated $248 billion in revenue in 2019. There are also numerous other companies which provide ground transportation facilities, services at the airports, food supply, and the various other services required to keep the system operational. The companies that operate are independent and exist to realize their own profits. Needless to say, the aviation industry is an excellent example of a large-scale, complex SoS. In this book, we will use the ATS as a recurring application example.

2.5 CHAPTER SUMMARY

In this chapter, we introduced the principles of Systems of Systems Engineering (SoSE). With the concept of *system*, we look at collections of elements as wholes rather than as a sum of discrete entities, and with *Systems Engineering (SE)*, we apply the concept of system as a basis for a structured approach for designing and building better systems to fulfill our needs. SoSE extends the ideas within SE for application to SoS.

1. Systems engineering (SE) is what results when we apply systems thinking for designing engineered systems. SE pays particular attention to clearly defining requirements early and meeting customer needs while involving stakeholders closely throughout the development process.
2. Modeling and Analysis (M&A) are useful activities undertaken in support of SE. Models are abstractions of the real world which include only those details of a system that are relevant to our task.
3. Models may be physical, analog, schematic, or mathematical. They can be categorized as descriptive vs. predictive, static vs. dynamic, deterministic vs. probabilistic.
4. SE and SoSE are intimately related, but with important distinctions that require a tailored approach for SoS development.
5. SoSE is the process of selecting the mix of constituent systems and designing their interactions. While SE focuses on developing independent complex systems, SoSE focuses on assembling the set of systems that would together provide the desired capabilities.

2.6 DISCUSSION QUESTIONS

1. Where have we seen failures in the verification and validation of systems in an SoS context? Research an example where verification and/or validation were not completed correctly. Provide two paragraphs describing how verification and/or validation were performed incorrectly and outline the resulting impacts.
2. Why is the "Engineering Vee" diagram used? What benefits does it offer for systems engineering? What are alternatives to this model in the SE community?
3. What are the benefits of, and challenges in, applying systems engineering in an SoS context, i.e. doing SoSE? Illustrate implications of both benefits and challenges encountered when extending SE to SoSE.
4. Define "Analysis" as described in the textbook for the SoS context.
5. How is working with computer models different than traditional scientific experimentation and inquiry?
6. Where is the greatest leverage applied in the SoS architecting process?
7. Chaotic systems are deterministic dynamic systems.
    a. True
    b. False
8. SoS is a "special kind of system." This fact is instantiated because:
    a. It is complex
    b. It is hierarchical
    c. Its components exhibit some kind of operational and managerial/control independence
9. While chaotic, non-linear, complex systems (like an airplane) may present design challenges, they are not generally "unmanageable."
    a. True
    b. False

3 A Formal Process of SoS Modeling and Analysis

In this chapter, we present a process for modeling and analysis specifically tailored to SoS problems with all their distinction introduced in Chapters 1 and 2. We start by describing key elements for proper SoS representation: hierarchy, lexicon, and taxonomy. Having a well-defined and commonly agreed upon lexicon is useful for enabling semantic interoperability among the various modelers, analysts, and decision-makers. Hierarchy and taxonomy are useful for relating components of the SoS into usable sets which can then be analyzed both in isolated and integrated contexts. Having outlined means for representing SoS, we then present a process for SoS modeling and analysis. This process is comprised of three phases: Definition, Abstraction, and Implementation (DAI). DAI is a structured approach to building models useful for evaluating and/or designing an SoS, regardless of its size or operational context. As with other chapters, illustrative examples from the air transportation domain explicitly show the rationale for why DAI works for SoS modeling and analysis.

3.1 SOS REPRESENTATION

3.1.1 SOS REPRESENTATION AND HIERARCHY

Effective SoS representation must contemplate the need for multiple abstraction levels each with the constituent elements and their interactions. Figure 3.1 illustrates the consequence of this seemingly simple statement! The interaction modalities, strength, and persistence may be quite different depending on the hierarchical level of representation. Further, representation of system elements at any particular level can take a variety of different forms including mere symbols on paper, variables defined in a computer program, or physical artifacts. The different forms of possible representation inform the different types of models which can be used for design, including mathematical models, physical models, logical models, etc. The process of developing system representations, or models, is called modeling. As described by Gell-Mann in *Quark and the Jaguar* [72], there is this kind of hierarchy of models in the physical sciences, starting with particle physics (quantum mechanical) as the most fundamental level, with chemistry sitting at the next most abstracted level, followed by biology, and then psychology, etc. Each successively higher level has more specific cases, exceptions, and rules that manifest the complexity we experience in our daily lives.

All science and engineering disciplines rely on some set of models of the system under examination (or development) to represent the analysis (or design) problem at

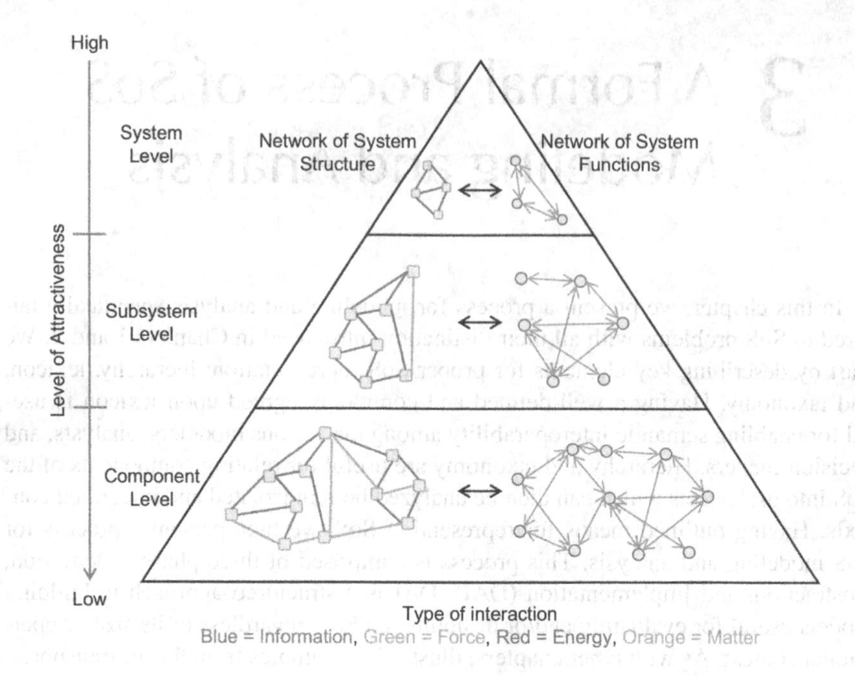

Figure 3.1: SoS is hierarchical, with emergent interactions at multiple levels.

hand, choosing the most appropriate hierarchical levels that best account for problem scope and purpose. The *INCOSE Systems Engineering Handbook* discusses different types of models such as physical, abstract, formal, and informal models [167]. Similarly, we need models to represent the systems engineering *process*. Classic examples of process models include the Vee-model and the waterfall model. In general, however, each discipline within a SE application uses a unique approach for representation, with its own syntax and semantics. This has prompted the SE domain to seek a common modeling environment, generally described under the banner of Model-based Systems Engineering.

Due to the relatively young age of SoS disciplines, representations of SoS problems, unlike traditional systems design problems, are not as advanced or standardized. We can mathematically represent some attributes of an SoS such as the constituent systems' autonomy, belonging, connectivity, and diversity, but such modeling becomes difficult as the scale and scope increase [14]. Furthermore, effects such as emergence may elude mathematical representation. Finally, as we will discuss later, SoS are a type of *socio-technical systems*, where modeling the social aspects is particularly challenging.

3.1.2 LEXICON

A lexicon is the vocabulary used for communication among those involved in a particular system or domain. A clearly defined and intuitive lexicon is essential not just for communication among stakeholders, but also for describing the unique nature of SoS. Ontology is a similar but distinct term. More than vocabulary, ontology is a set of logical relation between elements in a domain and (if a good ontology!) supports logical reasoning within the system. Ontology is important as systems analysis and engineering moves to model-based paradigms, but is not central to our goals in this text.

We seek a lexicon suitable for SoS problems that especially facilitates the description of hierarchy and heterogeneity in an SoS. It should clearly express the diversity of constituent subsystems at a suitable level of abstraction. Finally, it should be flexible to allow for representation of SoS of any size and type.

Table 3.1 shows a lexicon that meets the above requirements. This lexicon enables the description of any SoS along two major dimensions: categories of entities and levels of hierarchy, where entities in each category exhibit all of the levels. At the base, the α-level entities are the basic building blocks of the SoS; these are the lowest level systems that constitute the SoS possessing some degree of operational and managerial independence. A collection of α-level entities and their connectivity forms the β-level. Likewise, a collection of β-level entities and their connectivity constitutes the γ-level, and so on. While only four levels are shown, the total number of these can be increased or decreased as required by the desired scope of the particular SoS analysis.

Table 3.1
ROPE Table

Categories	Description
Resources	The entities (systems) that give physical manifestation to the SoS
Operations	The application of intent to direct the activity of physical and non-physical entities
Economics	The non-physical entities (stakeholders) that give intent to the SoS operation
Policy	The external forcing functions that impact the operation of physical and non-physical entities
Levels	**Description**
Alpha (α)	The base level of entities, for which further decomposition will not take place, α-level components can be thought of as building blocks
Beta (β)	Collections of α-level systems organized in a network
Gamma (γ)	Collections of β-level systems organized in a network
Delta (δ)	Collections of γ-level systems organized in a network

The four categories in this lexicon include Resources, Operations, Policies, and Economics, from which comes the acronym “ROPE.” Resources include all physical and non-physical components or systems that form part of the SoS. For example, aircraft and their communication links are resources in an Air Transportation System (ATS). Operations are the procedures and processes that form the dynamics of SoS. For example, all activities at the airports are operations in an ATS SoS. Policies are rules that govern management and operations. Limitations on number of flights during certain hours of the day, for example, is a policy in ATS. Finally, economics represent the needs and imperatives of stakeholders involved in the SoS, for example the revenue and expenses of airlines in the ATS.

Figure 3.2 is a visual representation of the four categories of resources, operations, policies, and economics, and four levels of hierarchy from the α-level of individual systems to the δ-level. The pyramid shape is used since the most number of elements exist at the lowest level, and then successive networks in the upper levels will naturally be lower in number. The ROPE table is a flexible framework which allows the various systems, contexts, hierarchy, and inter-relationships to be identified and described [46]. When used properly, and consistently, it helps teams avoid confusion, especially in multi- and trans-domain applications. It will be key in our exploration of complexity in SoS as well.

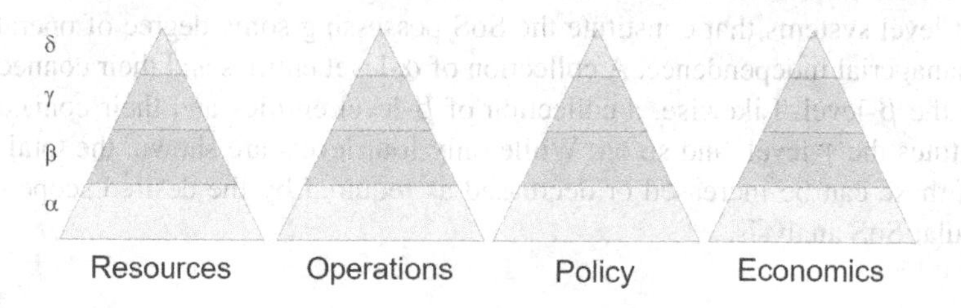

Figure 3.2: The "ROPE" categories and levels of hierarchy.

The ROPE table is particularly useful for identifying the resources, stakeholders, and context of the SoS. The ROPE construct is a structured approach to discovery and definition of variables. The hierarchical nature and sorting in categories favors sequential exploration of pertinent variables with respect to a traditional mass brainstorming. This discovery process mirrors the natural organizational hierarchies and structures often seen in both natural and man-made systems. Exploring a level of the ROPE table entails considerations on its definition and relations to the adjacent levels.

In the discussion that follows, we will use the four categories shown in the above ROPE table, although there is no stopping a designer from adding more categories as may seem useful for the particular problem at hand. Likewise, the number of levels of hierarchy and the size of each level would depend on the particular system under study. In some cases, the α-level will exhibit an order of magnitude larger number of elements than the β-level.

But there is an even more consequential observation to be made for SoS problems, one which is a key distinguishing feature of SoS as we discussed earlier. This is that the emergence property can only be perceived at the β-level or above. Thus, the behavior of the SoS is dominated by the structure and organization at higher levels as opposed to the characteristics of the α-level entities [45]. The implication of this, especially from a design perspective, is that the most consequential decisions are made at the higher levels of hierarchy. In other words, the question to answer becomes, "How does the preferred or observed behavior at the upper levels (e.g., γ-level) affect the possibilities for alternatives at the lower levels (α and β)?"

3.1.3 TAXONOMY

Taxonomy for SoS is a means to "divide systems of interest into two (or more) classes such that the members of each class share distinct attributes, and whose design, development, or operations pose distinct demands" [121]. Having a well-established and universally accepted method of differentiating an SoS from a conventional system would enable the development of tools and methods for the analysis and synthesis of SoS. For Maier [121], the main taxonomic node that distinguished SoS from 'not SoS' was the operational and managerial independence. But it is also useful to distinguish different instantiations within the SoS node.

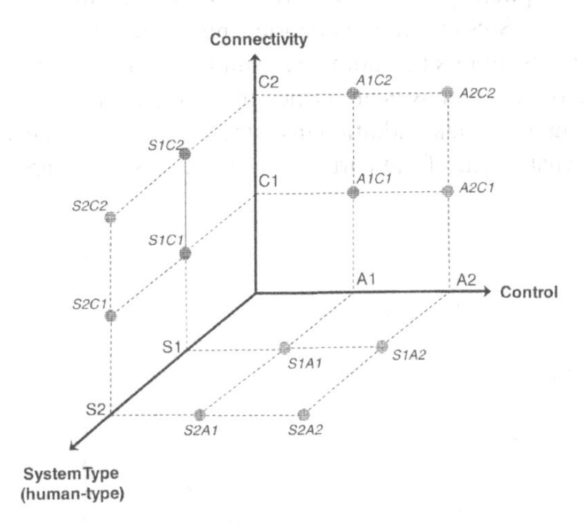

Figure 3.3: An example of an initial taxonomy: key dimensions and relative location of important SoS application domains [47].

One possible taxonomy for this within-SoS delineation is characterizing composition of SoS across three dimensions – system type, control, and connectivity [47] (Figure 3.3). The type dimension classifies systems along an axis with two extremes (it is possible to extend beyond two categories creating sets); the example shown in (Figure 3.3) implies a spectrum from largely technological vs. human entities. We noted earlier that SoS are a type of socio-technical systems. This means that humans are not just "external users and operators", but instead are most often actually part of the SoS as a type of physical entity. Unlike the classic monolithic engineered systems (e.g., airplane, rocket), humans have preferences, rational (and irrational) behavior, and the ability for participation in the system's function in a manner not completely under the SoS designer's control.

The control dimension describes the level of autonomy of elements at a level of hierarchy (or, alternately, the degree of centralized control) in an SoS. For example, most defense systems are at one extreme with a high level of central authority even though individual systems have operational independence, while the internet is at the other extreme with no centralized control whatsoever; the ATS, on the other hand, lies between these two. The final dimension is that of connectivity which is one key focus of SoS analysis (as evident in Figure 3.1). Along this dimension, we identify all the ways in which the constituent elements of an SoS interact with one another; do they exchange information, energy, or any physical mass? Do they collaborate or compete with one another during operation? This dimension, perhaps more directly than the other two in this example, taxonomy drives the emergent behavior indicative of a complex System of Systems character.

We can also characterize SoS via two inter-related means: boundary (endogenous, what is inside, and exogenous, what is outside) and nature (physical/explicit and non-physical/implicit) [45, 118]. Figure 3.4 shows that these two aspects divide the constituents of an SoS into four categories: resources, stakeholders, drivers, and disruptors. Resources, in this taxonomy, are similar to the resources category of the ROPE table. These are the physical entities that the stakeholders experience in an SoS; this makes them explicit. Additionally, since a system designer has full authority to decide on what entities form part of the system, resources are also classified as endogenous.

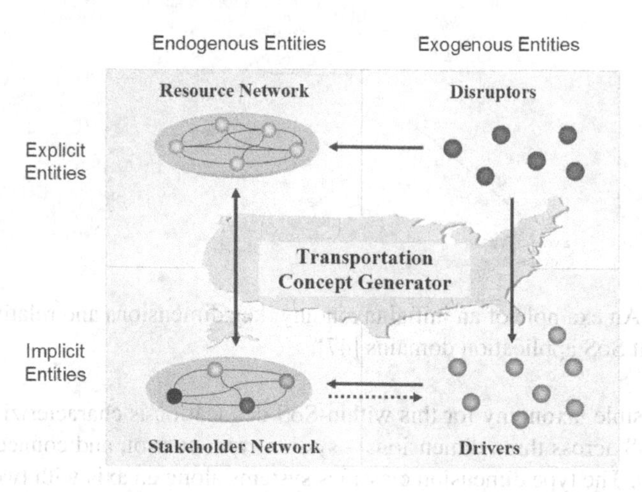

Figure 3.4: Classifying entities along explicit-implicit and endogenous-exogenous axes.

But one of the key characteristics of SoS is that they are not just comprised of physical artifacts designed and built to serve a purpose of the whole, but they also include a wider diversity of constituent elements, including humans and organizations, that have a self-interest in the operation and management of the SoS. This makes an SoS a socio-technical system. All of the entities that have an interest in the design and operation of an SoS are classified as stakeholders. Since not all of the stakeholders and their intentions may be known to all other stakeholders, this category of entities are classified as implicit. However, stakeholders still classify as endogenous because a designer does have some authority over who has a stake in the system, though the authority may only be partial.

Besides the previous two, there are two further classes of entities that are exogenous to a system architect's control. The first of these, driver entities are largely concerned with economic, societal, and psychological circumstances that influence the stakeholder network by implicit means. On the other hand, disruptor entities explicitly affect the resource network and/or a portion of the driver entities by reducing the efficiency of the resource network, disabling particular nodes or links of the network [118].

3.2 A 3-PHASE METHOD FOR SOS PROBLEMS

In this section, we will begin discussing a structured approach for modeling SoS problems. This approach is divided into three major steps, or phases, preceded by a preliminary scoping of the intended SoS evolution and general features of the SoS being evolved. A overview of the 3-phase method appears in Figure 3.5 and is termed the Definition-Abstraction-Implementation (DAI) method. DAI begins with a Definition (D) phase, where the purpose is to understand the SoS problem space including identifying the status quo and elucidating stakeholder requirements. These requirements may or may not be technical in nature, and they may encompass economic and policy considerations. In the second phase, called the Abstraction (A) phase, the task is to identify practical strategies to partition the SoS problem space, based on its scope and associated knowledge artifacts resulting from the Definition (D) phase, into relevant (and possibly non-intuitive) sets of resources, constituent entities, and their interaction networks. The phase includes consideration for practical constraints on both modeling and implementation perspectives. In the final Implementation (I) phase, solution frameworks (methods, processes, tools) are adopted to generate an end SoS solution. In the following sections, we will discuss each one of these phases in detail and illustrate the concepts using an example SoS problem of air transportation network design.

3.3 DEFINITION PHASE AND TOOLS

In the traditional approach to systems engineering, we would iteratively elucidate customer requirements and develop alternative solutions. In an SoS context, the very same task is beset by uncertainty driven by the operational and managerial independence of component systems. In fact, an SoS engineer likely would not have full (or even partial) control over the design of individual constituent systems but rather may exercise design authority at higher levels of SoS hierarchy. Thus, for SoS, defining

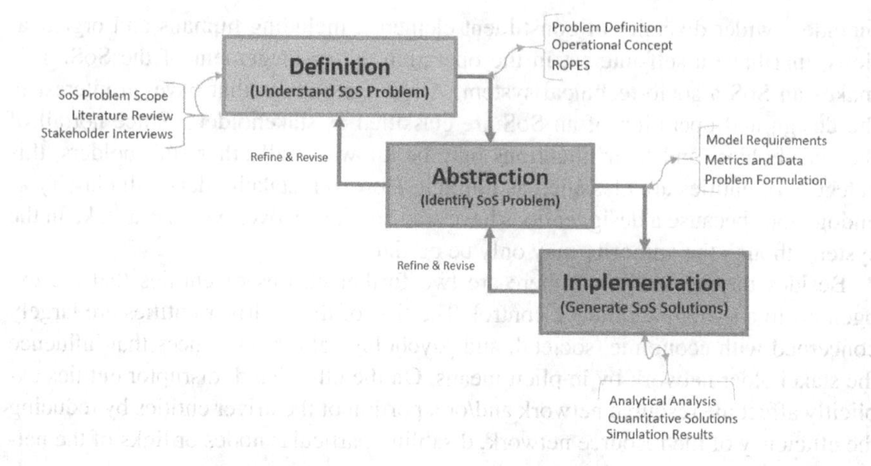

Figure 3.5: Overview of the DAI process.

boundaries, behaviors, and objectives is imperative before beginning any "real" requirement definition tasks. In general, the need is to understand the interdependence among individual systems rather than design those individual systems.

The primary task in the definition phase is to *understand* the problem including the context as well as the goals. The following are the objectives of this phase:

1. Identify the operational context including the domain, timescales, goals for new or evolved SoS
2. Describe the problem using the lexicon (scope and levels)
3. Identify barriers to preferred behavior in the status quo

Figure 3.6 provides a view of this phase as an input-process-output system. Notice that the "Definition" block contains the elements of the problem to be defined in this phase, viz., the operational context, status quo, barriers to obtaining the desired capabilities with existing systems, and the categories and levels in the ROPE table. We can now understand the utility of the ROPE table for identifying the entities within an SoS across its four categories and various levels of hierarchy. While there are many perspectives to an SoS, the activity of populating the ROPE table provides a structured, categorical, and sequential approach to guiding stakeholder explorations on constructing the SoS mental map. The discovery process, through the mapping activity, naturally gives rise to many issues on ensuring that an effective ROPE table is constructed to add value to subsequent steps in the DAI process. This table also serves as the primary mechanism to capture salient entities (endogenous-exogenous and implicit-explicit) related to each level of the SoS.

This being the beginning phase, the inputs include the SoS problem scope, and requirements which may be obtained, for example, from interviews with the stakeholders. Generally, we can expect that the process of requirements elucidation and problem scoping is an iterative process between the design team and the customer.

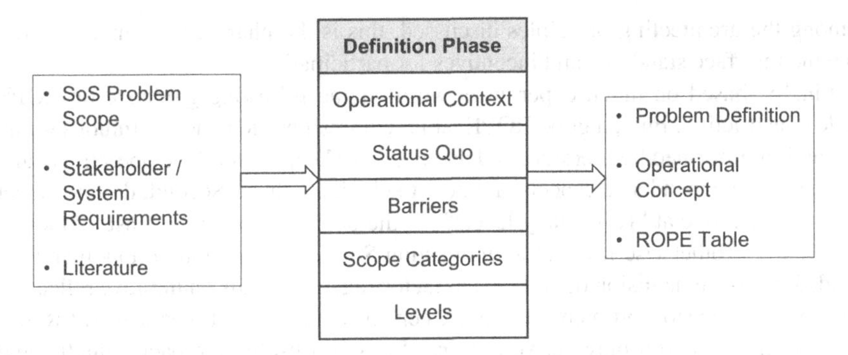

Figure 3.6: Definition phase diagram.

Existing literature, for example case studies pertaining to related large-scale systems design projects, can provide a baseline for comparison.

In summary, the objectives of this phase include understanding the SoS problem space, identifying key elements of an SoS, and describing these elements using a lexicon. Let us look at the process steps of the definition phase:

Operational Context: We identify the systems which are currently operational, the capabilities already provided, and the gaps that need to be further filled. Even if a subset of the constituent systems are to be newly developed, they will have to function alongside legacy systems which may be inflexible in their operation. Thus we need to identify the existing interfaces and operational paradigms.

Status Quo: We may not need to assemble an entirely new system to meet our needs. Sometimes, we may need to modify what is already deployed. We identify the status quo with regards to the currently functional systems and their capabilities.

Barriers: Do we need new systems to fulfill some needs? How much of the existing systems can be changed? Does the SoS systems engineer have the required control authority? Such are the questions to be asked when identifying the barriers to enabling systems.

Scope Categories: In this step, we begin to fill in the entities within the four categories of the ROPE table.

Levels: We identify not just the α-level systems, but also their networks, and networks-of-networks at the higher levels of hierarchy.

The systems engineering procedure analog to definition phase is the conceptual design phase. Early in the design process, the system is not very well defined and, hence, the designers have a lot of flexibility. Because of the uncertainty, the descriptions given in this phase can be qualitative. The outputs of this phase include, among other things, the ROPE table and a clearly defined problem statement. The challenge is to extend the existing guidance of the ROPE table to practically accommodate informational artifacts in a robust and organized fashion for real-world problems.

Among the architecting principles discussed, this is the phase where engineers define the interface standards and incentives for participation.

Finally, based on much experience, we offer the following guidance on ROPE table construction. First, a good ROPE table requires iteration and continuous group discussion, informed by literature and data. Always keep in mind the operational and managerial independence aspects and how to elucidate them. Second, do not expect to correctly arrive at "How many levels?" at the outset; certainly, at least two levels are required (otherwise it is not an interesting SoS problem) though how many are needed to ensure inclusion of all relevant factors again depends on iterative reflection on the core decision problem at hand. Keep in mind that at times a broad ROPE table from which you may derive multiple SoSs is a useful approach. Finally, and quite obviously, one cannot include all relevant descriptions in the cells of the ROPE table! Using the database/model of your choice, record richer data fields for each cell in a separate place that also may include parameter values, references, further elaboration/descriptions, etc.

3.3.1 EXAMPLE DEFINITION: ATS

In Section 2.4, we introduced the Air Transportation System (ATS) as an SoS. In this chapter and the following chapters, we will continue to use the ATS as an example to demonstrate various modeling approaches. Here, we will describe the definition phase of the DAI process as it can be applied to ATS.

There are three main types of planning decisions that airlines make: fleet selection, route planning, and schedule development decisions [17]. Each of these decisions directly impact the airline's profits and their share of the aviation market. These decisions have to be made with considerable uncertainty of market forecasts, while accounting for competitors' behaviors, and for several years in the future. References [130] and [131] discuss the development of a simulation tool used for the assessment of the environmental impact of aviation. This tool models an airline's aircraft allocation decisions made with the objective of maximizing their profit. The allocation model is part of a larger system dynamics-inspired approach for simulating the economics of airline operations, their decisions regarding retirement and acquisition of aircraft, and market demand growth.

To model these decisions, the complexity of the system needs to be managed; we will discuss the abstraction and implementation details in the following sections. Here, we first identify the operational context and the status quo. The above references restrict their models to the US air transportation network. The year 2005 is taken as the starting point of all simulations, which means that the state of the network and the airlines' fleet in that year is an input to the study. Market demand data and the average ticket prices were gathered. All the existing and yet to be modeled aircraft as of 2005 were modeled for inclusion into the simulated airline's fleet.

The entities included in the scope categories and across five levels of hierarchy for the Air Transportation System are shown in the ROPE table in Figure 3.7. The portion of the table included in the above references is shown highlighted in the figure. At the α-level, the above references model individual aircraft and their performance,

System-of-Systems Dimensions

Network of Networks

	Resources	Operations	Policy	Economics
α	Vehicles & Infrastructure (e.g. aircraft, runway, terminal area)	Operating a resource (e.g. aircraft, control tower)	Policies for single resource use (e.g. flight procedures)	Economics of building/operating/buying/ selling/leasing a single resource
β	Collection of resources for a common function (e.g. airport)	Operating resource networks for common function (e.g. airlines)	Policies concerning multi-vehicle use (e.g., local airport noise regulations)	Economics of building/operating/buying /selling/leasing resource networks
γ	Resources in a transport sector (e.g. air transportation system)	Operating a collection of resource networks (e.g. commercial air operations)	Policies concerning sectors using multiple vehicles (e.g. FAA fleet certification)	Economics of a business sector (e.g. commercial aviation)
δ	National Air Transportation System	National Passenger and Cargo flight movements etc.	National Air Transportation System policies	Forecasts of National Air Transportation Market
ε	Global Air Transportation System	Global Passenger and Cargo flight movements etc.	Bilateral agreements, ICAO regulations etc.	WTO; Global Marketplace

Figure 3.7: ROPE table for the FLEET project.

policies that govern their utilization, and their economics in form of operating costs. Besides aircraft, the studies also model airports, although at a much higher level of abstraction because no operational limitations are imposed on the airports. With reference to Figure 3.7, the above studies include levels α through δ; since these studies are for national air transportation network, the γ-level is not within the scope of the model.

3.4 ABSTRACTION PHASE AND TOOLS

Webster's Dictionary defines *abstraction* as "*a thought or view apart from material objects*." With the system and objectives clearly defined, the abstraction phase takes over with the following set of objectives:

1. Draw out the main classes of actors, effectors, disturbances, and interdependency networks
2. Focus on inter-relations among entities in a hierarchy, i.e., organizations and networks
3. Encapsulate the *big-picture dynamics* (not yet a detailed description)

In essence, the abstraction phase is a "bridging phase," facilitating the transition from the definition to the implementation phase. Via abstraction techniques, we identify the constituent entities of the SoS and their inter-relations and formulate relevant hypotheses about the problem at hand. We move from mostly words/descriptions in the definition phase to symbolic, logical, and graphical representations.

Outcomes of abstraction include the inputs, outputs, and metrics at each level of the SoS and a mapping of the interdependencies within and between hierarchical levels. This further clarifies the mapping of variables and the lexicon between different stakeholders. Going through a formalized abstraction process also helps us avoid the natural tendency to tailor an SoS problem formulation to fit pre-conceived implementation methods.

Thus, we approach the abstraction phase with the end in mind. We step away from the details and build the highest possible level of abstraction, and then we gradually add details until we are confident in building the right model. Sometimes abstraction leads to *architecture* problems and other times to *operations and control* problems. In particular, we focus on *three classes of design variables* in the abstraction process:

1. Composition – which systems/functions/resources?
2. Configuration – which operational interdependencies and constraints?
3. Control – what autonomy? Which incentives?

Composition pursues answers to questions such as, "Which systems should be present?," "What are the interface requirements?," and "How much will it cost, and at what level of risk?" These questions relate to a 'static' context. Configuration seeks to determine, "How are they connected?," "What information is passed on the links?," and "How does the connectivity evolve over time?" Control addresses, "How and who controls/influences the effectors?," and thus the latter two classes involve dynamics of the SoS. Once again, and referring to the ROPE table, these are resources, operations, policies, and economics, and, depending on the circumstances, some will be design variables, uncertainties, constraints, or objectives.

A key part of Abstraction is considering the timescale, or perhaps better put, the time horizon for the particular SoS problem. The *design timescale* – asking "Which SoS?" – is typically the longest and assumes an enduring SoS will exist. All the complications associated with long-time horizons exist here, and the open system nature of SoS adds further complications. The *planning and implementation time scale* – asking "How to assemble and prepare the SoS?" – is shorter, but still with potentially significant uncertainty on the constituent participants and their behavior. While shorter for component systems, it can be substantially longer for SoS due to continuous evolution of the SOS and the interdependencies between systems. Finally, the *operational time scale* – asking "How to Operate the SoS?" – is near-real-time with short decision windows; the multiplicity of perspectives and the need for quick decision synthesis to exploit opportunities for surprise synergy are most challenging here. In each setting, the design variable classes of composition, configuration, and control are all "active", just in different context.

In summary, the Abstraction (A) phase helps us achieve success in SoS modeling and analysis by bridging Definition (D) and Implementation (I). First, the abstraction phase helps circumvent a natural tendency to tailor an SoS problem formulation to fit pre-conceived implementation methods and solution approaches to siloed development cycles. The idea is that as we move towards modeling, we can avoid any *a priori* assumptions that may constrain the set of ideas that we may consider in developing a future SoS. As we model, proper abstraction guides us in gradually adding layers/complexity until we have appropriate fidelity to actually model an SoS. Second, in this phase, we elaborate on definition phase descriptions and build relationships by describing the various networks between the entities and developing entity descriptors along the dimensions of explicit-implicit and endogenous-exogenous. These entity descriptors are largely tied to the SoS problem scope (i.e., fully specified evolution, exploratory evolution, etc.) and guide the development of the SoS problem formulation and appropriately narrow the SoS solution space.

Thus, the entity-centric taxonomy discussed earlier is particularly suited to this phase of the process. We are interested in drawing out the major entities, actors, systems – technological or otherwise – the disturbances we may expect or not expect, and the interdependencies between these systems and the resultant network structures which represent them (Figure 3.8). The challenge that we seek to address with entity-centric abstraction (Figure 3.4) is to come up with the highest level of system representation that is "correct" for the SoS problem/design challenge under consideration, and, in the process, identify the following in the abstraction phase:

Stakeholders: We begin by identifying the stakeholders and their needs. Stakeholders are considered implicit because you cannot go "touch" these entities; they may not be tangible.

Resources: We identify the resources that are the tangible systems which form the SoS. These are explicit endogenous entity under control of the SoS designer.

Drivers: These are those entities that determine the forcing functions that drive the stakeholder network and receive a feedback from the stakeholders.

Disruptors: These are explicit exogenous entities which directly affect the resource network. These entities have an indirect effect on the stakeholder needs.

Networks: Ultimately, it is the network of resources through which an SoS gets its capabilities and behaviors. The behaviors in resource and stakeholder networks influence each other.

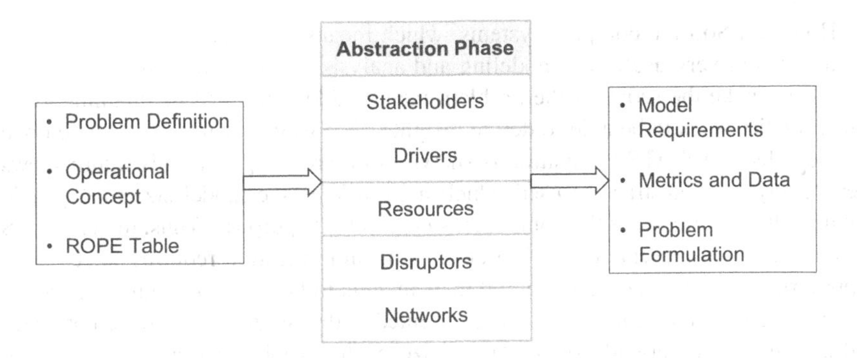

Figure 3.8: Abstraction phase diagram.

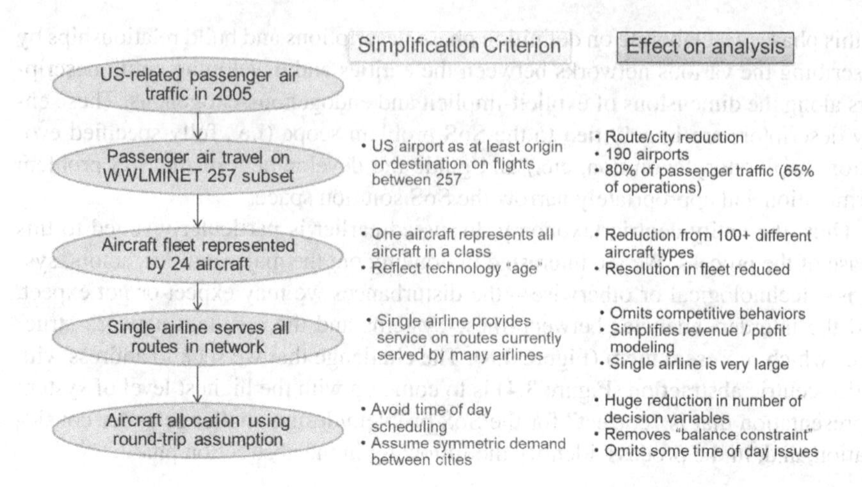

Figure 3.9: Model abstraction for the FLEET project.

3.4.1 EXAMPLE ABSTRACTION: ATS

The ROPE table displayed in Figure 3.7 identified the resources at various levels of hierarchy for an air transportation network. Similarly, some of the stakeholders, drivers, and disruptors can be identified from this table. For example, vehicles and infrastructure such as airports are the resources which constitute the system. Entities such as airlines, regulatory agencies, and passengers are stakeholders. Finally, phenomena such as economic conditions, the weather, or even policies can be either drivers or disruptors depending on the role they play and the objectives of the various stakeholders. These can then be laid out in the quadrants of entity-centric abstraction diagram.

However, SoS are complex systems, which means that merely identifying their entities is not very useful for modeling and analysis. We need to abstract away the details to make the scope of the problem manageable while still maintaining sufficient fidelity so as to be able to derive insights. One way in which this can be done for our Fleet-level ATS Evaluation is shown in Figure 3.9 [131]. This figure shows several layers of abstraction used which help to keep the model size manageable while still accounting for the components crucial to its purpose. Thus, the entire US airline fleet is represented by 24 aircraft. This simplification requires selection of appropriate set of 24 aircraft and and their models to be used for simulation. Similarly, all airlines are aggregated and represented with a single airline which operates all aircraft in the network, reducing the size of the problem but at the expense of excluding effects of competition. Elimination of competition requires modeling the airline to be a "benevolent monopoly," which means that the airline does not exploit it monopoly status by increasing ticket prices. Instead, ticket pricing is based on a fixed margin of profit that the airline earns from every passenger it carries.

3.5 IMPLEMENTATION PHASE AND TOOLS

The implementation phase is the "culmination phase," which completes the purpose of modeling and analysis, but can lead to an iteration back to the definition phase for updates or verification. It is in this phase that we instantiate the aforementioned conceptual abstraction of the SoS via simulation to try to see if we can find emergent behavior. Given the complexity of the problem at hand, which makes both experimentation and a search of analytical solution an ill-advised endeavor, this phase is the primary focus of our modeling efforts. The objectives in the implementation phase are:

1. Instantiate all or part of the abstraction within a computer model housing *appropriate* methods/algorithms
2. Test, verify, and validate the model
3. Propose and test specific hypotheses about the SoS *to guide decision-making*

Thus, in this phase we generate SoS solutions via applicable methods and balance the complexity of a possible solution with technical and programmatic risk. To do so, we define the elements of the model used for generating solutions, including objects to be included in the model, the methods which define the objects' behaviors, and even the data exchanges involved (Figure 3.10).

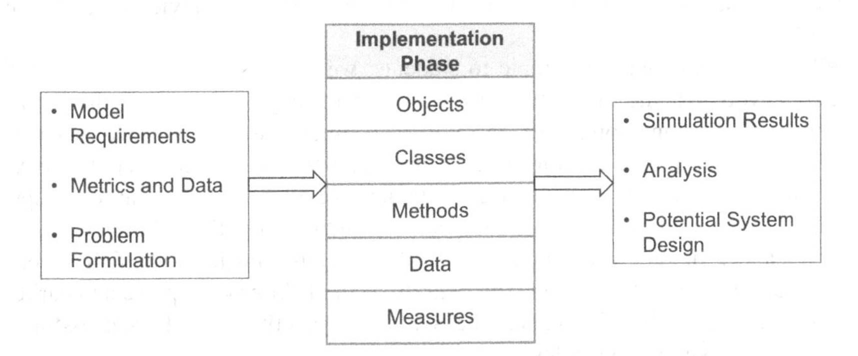

Figure 3.10: Implementation phase diagram.

We need to define the following features of a solution approach in this phase:

Objects: Objects are constructs that contain their data and methods and execute these methods whenever they are called for in a simulation exercise. In our terminology, when we identify the resources and the actors constituting an SoS, we identify objects to be defined during implementation.

Classes: These are the generic abstractions of the objects that we have identified earlier. These encode the recurring patterns of structure or behavior that can be used to classify the constituents of a system.

Methods: Due to heterogeneity of the system, we seek to allocate an appropriate solution strategy to subdomains in the SoS involved for an overall analytical tractability. Thus, we select the *models* for our modeling and analysis, including the formulation of our problem statement, and algorithms for solving our problem.

Data: During analysis of SoS, simulation takes place of "true" experimentation which is unlikely to be feasible for any design tasks. Data in form of historical records or that obtained from simulation is desired if we hope to derive any insights on such complex and non-linear systems as an SoS.

Measures: While the measures of performance from SE can be applied to each of the resources, to capture higher-level features of an SoS, particularly emergence, we need novel measures. The tools, methods, and metrics will depend on the specific SoS problem, the objective of the analysis, and the formulation produced by the abstraction phase.

3.5.1 VERIFICATION AND VALIDATION

In nearly all circumstances, an SoS M&A activity produces outcomes that must be convincing to some person or organization who did not do the modeling! Thus, the universally important activity of Verification and Validation (V&V) is integral to complete the implementation phase (and iteration through multiple turns of the DAI process). The general, and disciplinary specific, body of knowledge on V&V is vast. In this short section, we present just a few fundamentals with advice on their use tailored to SoS.

When first introducing the topic to students, we emphasize that V&V is NOT something you do at the end! V&V must start from the beginning, even in the definition phase. Our second point of emphasis is to encourage budding SoS M&A practitioners to "keep it simple", following good modeling practices from the start to maximize chances for success at the end. In particular, we offer the following four steps, gleaned (with small tweaks) from a short course conducted by Richard Meltzer [127] and colleagues at the New England Complex Systems Institute (NECSI), well-encapsulate these good practices. The entirety of the DAI process presented earlier in this chapter is designed to enable the SoS M&A practitioner to be successful in following these simple four rules:

1. Build model with minimal complexity to reproduce key behavior.
2. Use minimal set of parameters.
3. Test thoroughly, continuously, to understand logic of what is going on in the model and develop stable intermediate forms.
4. Add complexity (M&S functionality) gradually, as necessary.

With this simple concept on the table, we next introduce students to a more rigorous model understanding (and executing) V&V, with tailored insights for the SoS concept. The model we present is from the well-known work of Robert Sargent [146]. The following is an SoS-oriented elaboration model, centered around understanding our view of the model (depicted in Figure 3.11) inspired by Sargent's original conception.

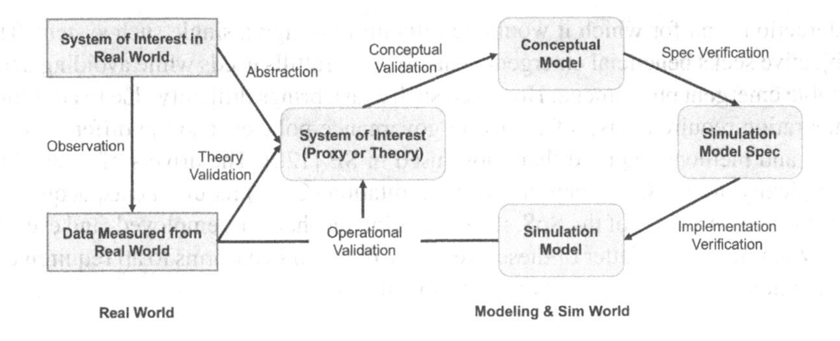

Figure 3.11: Recasting verification and validation (after Sargent) for SoS M&A.

We often think of the SoS under study as it currently exists (or we imagine it might) in the real world. But in fact through the definition and abstraction phases, we diligently work to create some proxy model (or a theory, and/or a set of equations, etc.) of the SoS. Together with specific research questions we seek to answer, this produces our conceptual model which emanates most directly from the abstraction phase. So, conceptual validation is achieved if we do these steps well and can produce evidence for congruence (with traceability) between the SoS in the real world and this conceptual model. Next, we derive specifications for the computer model we intend to build and then build it (hopefully applying steps 1-4 listed above). We call these two activities "verification steps"; verification relates not to whether our model well-reflects the system of interest (that is validation), but whether we are being faithful to the conceptual model in terms of the computer model/simulation developed in the implementation phase. Finally, we obtain operational validation if the combination of data from the real world and outputs of the simulation together adequately represent the system of interest for the *specific hypotheses* being tested (and not every conceivable hypothesis!). Testing our proxy model/primary theory of the system of interest directly to observed data from the real world would be an exercise in theory validation.

A few words about new (and old) paradigms pertaining to industrial applications of systems/software development and V&V. First, the move to agile development, continuous integration, etc. in software and enterprise-wide Digital Engineering (DE) in more general systems development has, is, and will transform V&V of modeling and simulation. As complexity of humanity's creations inevitably increases (with SoS integration as a prime example!), such advances are critical to perform both verification ("was the product built right?") and validation ("was the right product built?"). Further, these and other related approaches promise to improve quality attributes beyond V&V, such as model accreditation and certification. The following scenario both illustrates this and recaps some of the most important aspects that characterize SoS.

During integration of multiple systems in an SoS context, the objective is to bring together a collection of systems with capabilities that arise from the systems' interactions and for which it would be difficult to design a single such system. This objective seeks beneficial emergent behavior, that fulfills needs while avoiding undesirable emergent phenomena. However, such a task brings difficulty, due to additional integration requirements, difference in governance policies, and the different set of tools and methods required than those used in SE [120]. The drivers of such added complexity are the Certification and Accreditation (C&A) requirements, acquisition approaches, structure of the SoS, the integration mechanisms employed, and even the V&V applied. In the latter of these, we verify if the SoS conforms to all requirements and validate if the SoS has all desired capabilities.

3.5.2 EXAMPLE IMPLEMENTATION: ATS

The implementation method(s) employed in modeling our Fleet-level ATS Evaluation example must flow from the definition phase, especially the specific research question under study, and the abstraction phase, especially the model scope/requirements and family of hypotheses. We will treat the important task of defining appropriate research questions (and hypotheses) at more depth in the next chapter. The FLEET project made use of a system dynamics-inspired framework that mimics the economics of airline operations, models their decisions regarding retirement and acquisition of aircraft, and estimates market demand growth. The central problem of airline allocation represents the airlines' profit-seeking operational decisions as a mixed-integer programming problem.

Notice how in Figure 3.12, which shows this implementation, we can see the interactions between the different components of this problem. For example, we see that not only the fuel costs, but also the taxes levied on fuel contribute to the total operating costs of the airline. Or that the number of aircraft delivered to the airline depends not only on the number of retirements, which is a decision dependent on the operating economics of a particular aircraft, but also on the available number of aircraft due to the limit on number manufactured.

In this figure, the text that is italicized indicates inputs to the tool, for example, the Entry-in-Service (EIS) dates of aircraft. All other blocks are models. The central block, Airline Fleet Allocation, is where the airline's decision-making takes place, and is a good case study on the kinds of decisions that implementation phase requires. The mixed integer programming problem of airline's decision-making uses profit maximization as its objective function. Constraints in this problem include those on operating hours of the aircraft, the seat capacity within an aircraft, and on the minimum demand served by the airline on a route. So far, in this chapter, we have done a brief discussion of FLEET project as an example for all three phases of SoS modeling and analysis; for further details on modeling, refer to [131].

3.6 CHAPTER SUMMARY

Modeling and analysis to support the design, development, and planning of a System of Systems (SoS) (and its desired) evolution is a difficult task! This chapter has identified some reasons for this including the vast number of (potential heterogeneous)

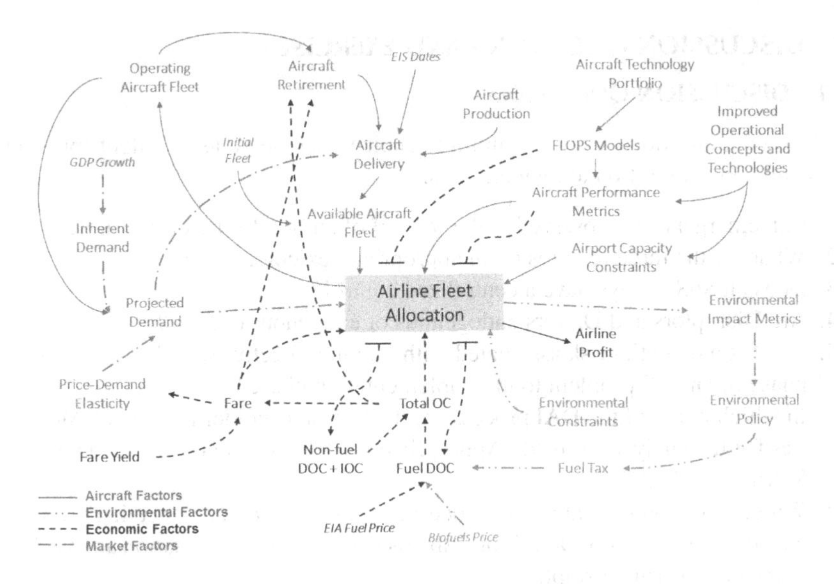

Figure 3.12: System dynamics representation of FLEET.

systems involved, multitude of organizational incentive structures, compartmentalization of information, and, complex behavioral dynamics across levels in SoS hierarchy. Each of these factors, among others, makes it difficult to comprehend, model, and subsequently, synthesize practical solutions. Motivated by these facts, this chapter introduced the three-phase Definition, Abstraction, and Implementation (DAI) process. DAI provides an application domain-agnostic approach for SoS modeling and analysis incorporating both qualitative reasoning and quantitative analytics to create knowledge artifacts tailored to decision-support in SoS context.

1. The definition phase is about understanding the problem. We define the operational context and status quo, barriers to implementation of new systems, and the ROPE table which includes the scope of systems and their networks organized into levels of hierarchy.
2. The abstraction phase is the bridging phase between problem definition and solution implementation during which we identify the resources and stakeholders and their networks, and exogenous entities of drivers and disruptors of the SoS.
3. The implementation phase is when we instantiate the model and simulate. In this phase, we define system classes and objects, and set up the methods that enable their functions. We also define the data and measures of performance associated with the SoS.

3.7 DISCUSSION QUESTIONS AND EXERCISES

3.7.1 DISCUSSION QUESTIONS

The following questions, even the multiple choice ones, serve as excellent spurs for discussion and exploration among learners.

1. Can emergence be observed at the α level of an SoS? Why or why not?
2. What are the three features of an appropriate lexicon for SoS?
3. Does an SoS always have a centralized control?
4. Are Disruptors and Drivers endogenous or exogenous entities?
5. What is a potential risk associated with moving directly from the definition phase of an SoS problem to the implementation phase?
6. In which phase of the DAI process are the SoS stakeholder identified? Metrics for the analysis defined? Appropriate models identified and discussed? ROPE table built?
7. Where have you seen DAI-like processes implemented in engineering? Discuss examples of how Definition, Abstraction, and Implementation are used in the engineering disciplines.
8. Context and boundary are important for defining (and modeling) a system.
   a. True
   b. False
9. Is it difficult to solve many SoS problem effectively by manipulating elements ONLY at the α-level (i.e., lowest level)?
   a. Yes
   b. No
10. How many hierarchical levels are required in a ROPE table?
   a. 4
   b. 2
   c. 6
   d. It depends on the particular problem/scope
11. Which of the following is NOT a source of complexity in an SoS?
   a. Uncertainty of requirements
   b. Multiplicity of perspectives among stakeholders
   c. Lack of diversity among components
   d. Unmodeled interdependencies
12. Which of the following is NOT the barrier to successful SoS modeling and analysis?
   a. Tendency of fragmented perspectives for practical development
   b. Lack of mathematical understanding of physical systems
   c. Lack of theory or methodology for integrated systems analysis
13. Which of the following would you NOT be concerned with during the abstraction phase of the DAI SoS M&A process?
   a. Specifying the main classes of actors
   b. Recognizing the "big picture" dynamics
   c. Coding the exact behavior of the agents
   d. Identifying the networks that link the entities

14. What are the ROPE table categories?
    a. Research, Operations, Project, Economics
    b. Resources, Operations, Politics, Environment
    c. Resources, Operations, Policy, Economics
    d. Reusability, Operations, Policy, Energy
15. Which is NOT a class of SoS design variables?
    a. Composition
    b. Configuration
    c. Control
    d. Communication

3.7.2 EXERCISES

1. Design a system hierarchy for a real-world System of Systems problem. Perform background research on the different levels of system hierarchies and decide on an application of interest. Discuss why you chose to place specific aspects of the system at particular levels.
2. Think of an SoS problem of interest. With the help of the example provided in this chapter, try to fill a ROPE table for the specific problem. Do not focus only on the Resources (component systems), but on all the aspects that will impact the problem.

Part II

Methods and Tools for System of Systems Modeling and Analysis

4 Bridging Theory and Practice

4.1 CHOOSING THE RIGHT QUESTIONS

In the Preface, we advised all to heed the wisdom of Richard Bellman, who stated that, "The right problem is always so much harder than a good solution." In Chapter 1, a history of systems modeling via the "Solberg Chart", indicated the immense diversity of modeling methods with accompanying opportunities and challenges. The DAI process elaborated in Chapter 3 is our practical approach to following Bellman's guidance, especially in the implementation phase where tools must be selected (or created) *and validated* to address the properly posed research question (RQ).

While the term "Research Question" is common usage in academic settings, our plain meaning is the essential question to be answered and/or decision to be made regarding a particular SoS problem. For example, in the healthcare industry, a RQ might center on the proper connectivity of distributed medical monitoring devices across a network of hospitals. In transportation, a RQ may address the appropriate degree of autonomy in a fleet of package delivery vehicles that increase overall system efficiency while respecting current (and emerging) operational policies. A non-trivial barrier, however, is that the trans-domain nature of SoS means that one may need to integrate several distinct fields of inquiry, each with its own style and requirements, for research question development.

Many high quality references exist on developing good research questions, far too many to cite here and mostly tailored to a particular domain. However, one broadly applicable source we recommend is Creswell [37]. While stemming from the social science realm, this work deals with both quantitative, non-quantitative and mixed methods and thus is a useful general treatment of most SoS research questions and hypotheses. As the title of Creswell's book clearly suggests, there are qualitative, quantitative, and mixed methods research design approaches and each has both common and unique elements. Qualitative research "is an approach for exploring and understanding meaning individuals or groups ascribe to a social of human problem," quantitative "is an approach for testing objective theories by examining relations among variables," and mixed methods "is an approach to inquiry involving collecting both quantitative and qualitative data, integrating the two forms, and using distinct designs that may involve philosophical assumption and theoretical frameworks." SoS M&A settings will work in the quantitative methods domain for the most part, though attention to mixed methods is not unwarranted. For example, research in the humanities and social sciences on how organizations behave and evolve may be very important in understanding manifestation of operational and managerial independence in SoS. In the course of both our research experiences and teaching the AAE 560 course, a number of cases have arisen requiring instruments other than physics-based modeling or experiments, including especially surveys and assessments of policy regimes, etc. See Section 4.4, which provides a long list of project titles, including several which clearly exemplify this.

Several fundamental elements of Quantitative RQs and Hypotheses are summarized by [37]. A quantitative RQ asks about the relation among variables, while a quantitative hypothesis makes predictions on outcomes involving these variables. Engineering students are (or should be) familiar with this especially as they enter the last year of undergraduate studies or for sure by the time they enter graduate school. A good chemistry course or statistics course, for example, should demand this of them. First, the independent and dependent variables must be measured separately; as we will see in Chapter 6 on agent-based modeling, this is no trivial task as model complexity increases. Also, students should be aware of the two types of hypotheses available. A *null* hypothesis proposes that no difference (or relationship) exists between the variables under study, while a "directive" hypothesis involves a prediction about an expected outcome, e.g., some variable will be lower (or higher) as a function of increasing value of another variable. Finally, this work highlights that the most rigorous form of RQ/Hypothesis is testing a theory and in such cases the variables involved must be examined in appropriate relation to their role in the theory. This is important in SoS as the Modeling & Analysis method selected is usually more than just an input-output correlation exercise. The essence of the theory underlying the method is central to the expected SoS behavior under study. Thus, the RQs and Hypotheses must respect (and exploit) this theory for explaining expected behavior and justifying suggested design choices.

Let us next summarize in a non-formal way some guidance on developing RQs and associated hypotheses "from the field", i.e., from our experience in teaching SoS courses. We begin by offering (with tongue in cheek) the following question: what is a valid RQ? The answer is: something that can be researched, under constraints (time, resources, etc.) and producing a well-scoped finding, enabling decisions which somebody cares about (e.g., an SoS manager, a system owner debating whether to join an SoS, a policy-maker seeking long-term sustainability, etc.). Such a simple explanation provides a nice complement to the rigorous definitions we summarized in the previous paragraphs.

We then offer a bit more detail to enhance the general guidance on developing good RQs and hypotheses. Further, we make clear that this additional detail forms the rubric for evaluating Definition and Abstraction Phase products. Despite their simplicity, students are frequently amazed to see that their initial efforts obviously violate some of these guidance items (a sign that it is not as easy as it looks). Listed next are these detailed good properties of RQs and Hypotheses:

1. "Goldilocks" – specificity: Not too broad, not too narrow; use your ROPE table to isolate the levels and categories relevant to your problem.
2. Doable: Can be researched (including problem formulation, modeling, analysis) in given amount of time/budget/data availability.

3. Translatable and actionable for a *specific* person/role, groups of people, organization.
4. Aspirational: Generates a better outcome than the status quo: a better method to get a acceptable outcome/answer; or a method to get a better outcome/answer (like an improved SoS architecture).
5. Consistent/compatible with other DAI elements, especially ROPE table; a good RQ should be clearly understood while viewing the ROPE Table.

In light of these properties, especially the first one on specificity, SoS M&A practitioners should be on the look-out for telltale signs of imprecision, such as: “It is hoped that a basic understanding...” (too vague); “Hope to model some complexities...” (too vague and broad); “How would improved communication enable mission success...” (too broad); “Seek more efficient operations” (too vague). While SoS is “aspirational,” we do not “hope” for things in research questions. Table 4.1 gives three examples of “good work” (nothing is ever perfect) from past AAE 560 SoS M&A team projects. Each RQ has clear ties to categories of the ROPE and/or addresses one or more of the three classes of SoS design variables (Section 3.4). For example, the first RQ posed by the project addressing water rights in Peru centered on the Economics and Policy category of the ROPE table, with emphasis on the γ level with contributing elements from the beta and alpha levels. Each hypothesis was tied to the RQ, testable via simulation using specific design variables and metrics. For example, the PlexNET project built an SoS model whose SoS model produced strong evidence for their hypothesized effect of vehicle heterogeneity and information exchange on exploration value. In fact, this project developed into a technical publication (as quite a few have over the years), presented at the 2020 AIAA SciTech Forum [158].

Let us now turn to this topic of model/method selection, revisiting the important aspects of the DAI’s implementation phase through the lens of the wonderful examples of method and theory in the “Solberg Chart.”

4.2 CHOOSING THE RIGHT TOOLS

Each of the many methods illustrated in the Solberg Chart (Figure 1.1) carries a deep history and evolving applicability for SoS. The DAI process is the key to selecting the right method for a particular problem. However, some rudimentary knowledge of the method is necessary, and therefore it is appropriate to examine some of these methods in at least slightly more detail to include a specific reflection on their applicability to SoS according to reflection of the distinguishing traits of SoS as well as the three key classes of SoS design variables (composition, configuration, control) introduced in Section 3.4. The following section contains brief summaries of the outcome of this examination. There is a detailed exercise recommended for this purpose offered at the end of Chapter 1 for instructors using this text in a course. However, the exercise in exploring the Solberg Chart in more depth is a worthwhile activity for anyone seeking to practice the art of choosing a modeling method for Implementation based on solid Definition and Implementation phase outcomes.

Table 4.1
Real Examples of Good Project RQs and Hypotheses

Topic	RQ	Hypothesis
System of Systems Supply Chain Study with Lunar Gateway and Lunar Colonization Base	What composition (i.e. types and number of subsystems) is required in the Lunar Gateway, lunar ground-base, and other Artemis systems in order to minimize the cost of resupply and maintenance?	Lunar production of critical resources will significantly increase the robustness of the supply chain SoS.
Water Rights and their Management in Peru	How can Peru restructure its financial system to minimize water conflict? What is the optimal path of water distribution on a regional basis?	Peru's government can create net benefit to regional stability and economic health by reducing mining operations in the country via reduction in conflict with local populations over water rights.
Semi-Autonomous Planetary Exploration Network (PlexNET)	What configuration features in a distributed SoS exploration approach maximize effectiveness increase over the status quo (single monolithic vehicles per mission)?	Heterogeneity in vehicle squadrons and information sharing among vehicles generates feasible architectures that produces higher exploration value per unit cost.

Adaptive Systems

Founder/Early Pioneers. Drenick and Shahbender, 1957.

Underlying principles. Parameter estimation, Lyapunov stability, sampling, gradient methods, and Barbalat's lemma.

Well-known applications. Flight control, autonomous driving, robot machining arms, and perturbation dampers.

Limitations. Adaptive systems are highly domain-specific, prompting unpredictability outside of specific implementations.

General applicability to SoS problems. Adaptive systems exemplify the operational independence of systems in an SoS construct.

Ant Colonies

Founder/Early Pioneers. Marco Dorigo, 1992.

Underlying principles. Optimization and general pattern recognition techniques.

Well-known applications. Ant colony studies, crowd and mob behavior analysis, and traveling salesman optimization.

Limitations. Large models require significant time and computational power. Size and scope of models are primary concerns of implementation.

General applicability to SoS problems. Autonomous swarm-like simulations with operational and managerial independence represent SoS architectures.

Artificial Intelligence

Founder/Early Pioneers. John McCarthy and Alan Turing, 1950s.

Underlying principles. Probabilistic methods, search and optimization, and biological networks and reasoning.

Well-known applications. Robotics, transportation, mobile and web applications, healthcare, image and speech recognition, and games.

Limitations. Reasoning for decisions and purpose - unawareness. Processing unseen inputs could result in emergent behavior. Limited learning capabilities due to data and computational capabilities.

General applicability to SoS problems. AI performance meets the needs of automation and complexity in SoS (e.g., integrated defense networks, smart cities).

Artificial Life

Founder/Early Pioneers. John Von Neumann, 1986.

Underlying principles. Biology and automata theory.

Well-known applications. Wetware systems, artificial chemistry, and white-box modeling.

Limitations. Certain behaviors cannot be replicated through machine specification or coding, either due to theory or a lack of knowledge.

General applicability to SoS problems. Main tenets, inspired by living organisms, mimic features of SoS construction and behavior (e.g., evolutionary behaviors, emergent properties, etc.).

Automata Theory

Founder/Early Pioneers. Alan Turing, 1936.

Underlying principles. Mathematics of Turing machines.

Well-known applications. Computer science, linguistics, evolution, mutation, and natural selection.

Limitations. Finite state automata models only work with singular variables at a time. Turing machines only run a single instruction set.

General applicability to SoS problems. Connections to various components or systems are modeled through automata theory. Enables modeling of virtual SoS networks.

Biological Learning and Intelligence

Founder/Early Pioneers. Many.

Underlying principles. Evolutionary biology, psychology, and genetics.

Well-known applications. Genetic influences, immune system, central nervous system, and psychological influences limit scope and physical possibilities within biological systems.

Limitations. Socio-economics and environment inhibit modeling of biological intelligence. Difficult to distinguish between natural biological learning vs. influenced learning.

General applicability to SoS problems. At the system level, biological learning is an example of Directed SoS. At higher levels, it can represent Collaborative SoS (e.g., vaccine distribution across a diverse demographic).

Catastrophe Theory

Founder/Early Pioneers. Renee Thom, 1960s.

Underlying principles. 3D surface modeling and multivariate systems.

Well-known applications. Stock market volatility, structural loading/failures, and physio-chemical processes.

Limitations. Impossible to visualize models due to multi-dimensional spaces. Significant over-simplification of complex systems.

General applicability to SoS problems. Interfaces between two systems and overall network robustness can be tested by observing control variables defined in catastrophe theory.

Cellular Automata

Founder/Early Pioneers. Stanislaw Ulam and John Von Neumann, 1948.

Underlying principles. Automata theory and graph theory.

Well-known applications. Conway's game of life.

Limitations. Exponentially scaling computational requirements, simulation limited to discrete timesteps and automata states.

General applicability to SoS problems. Computational modeling for agent-based models allows swarm-like SoS to be modeled with well-defined objects and agents.

Chaos Theory

Founder/Early Pioneers. Edward Lorenz, 1961.

Underlying principles. Causality and determinism.

Well-known applications. Weather modeling, plasma physics, economics, and 'butterfly effect' models.

Limitations. Starting conditions are assumed to be perturbation-sensitive. Precise understanding of initial state required to obtain results.

General applicability to SoS problems. Evolution behavior and chaos within an SoS can be described, provided that the bounds of the stochasticity in the model are known.

Complexity

Founder/Early Pioneers. Many.
Underlying principles. Distributed control and connectivity.
Well-known applications. Genetics, biological systems, social analysis, and communication networks.
Limitations. Adequate system boundaries are difficult to define. Emergent behavior is difficult to find/track in models.
General applicability to SoS problems. Applied directly to emergent behavior and evolution, key traits of an SoS.

Computer Supported Cooperative Work

Founder/Early Pioneers. Paul Cashman and Ierene Greif, 1984.
Underlying principles. Computing, mathematical logic, and networks.
Well-known applications. E-mail, version control, computer networks, and social media.
Limitations. Limited human proficiency with computers, lack of standardized internet access, and information security are physical/regional limitations for interconnection.
General applicability to SoS problems. Helps define the interfaces between human-controlled systems and autonomous systems in an SoS.

Control Theory

Founder/Early Pioneers. Aristotle, 3rd century BC.
Underlying principles. Calculus, mathematical logic, and programming.
Well-known applications. Almost everything! From thermostats to airplanes, transportation systems to robotics, and everything in between.
Limitations. Functions must be independently programmed. High physical/computational resource utilization if problem scope is large/complex relative to existing technologies and human abilities for implementation.
General applicability to SoS problems. Can be used to design regulators in SoS networks. Helps determine impact of perturbations within and between systems.

Cybernetics

Founder/Early Pioneers. Norbert Weiner, 1948.
Underlying principles. OODA loop, objective attainment, and control systems.
Well-known applications. "Governor" for controllers, feedback response systems, and automated flight controls.
Limitations. Cybernetics reveals needs without solutions. Difficult to view/measure data for an SoS.
General applicability to SoS problems. Complements Maier's operational and managerial independence distinguishing traits in virtual SoS architectures. Applicable to virtual SoS.

Discrete Event Simulation

Founder/Early Pioneers. Keith Torcher, late 1950s.

Underlying principles. Management of entities, queues, events, and resources.

Well-known applications. Industrial process or chains, warehousing, queues, road traffic, business process management, operational security, healthcare service.

Limitations. DES is not a desired method for continuous and dynamic operations. Hierarchical modeling techniques cannot be included, so DES is not suitable for a complex multi-level system.

General applicability to SoS problems. Useful in cases where interactions between "independent" systems of an SoS are asynchronous and event-triggered by nature.

Economics

Founder/Early Pioneers. Adam Smith 1750-80s.

Underlying principles. Integral calculus, matrix algebra, and computational methods.

Well-known applications. Market optimization, static and dynamic analysis of economic units, comparative statistics, and commodity analysis.

Limitations. Cannot account for human behavior when modeling economic systems. Models cannot cope with variation due to unpredictability of human behavior.

General applicability to SoS problems. One of the key elements to defining and designing SoS is the economics of the architecture and its constituents. This allows for optimization of assets (e.g., airline fleet management).

Emergence

Founder/Early Pioneers. George Henry Lewes, 1875.

Underlying principles. Coherence, dynamics, ostensivity, and radical novelty.

Well-known applications. Organizational dynamics, computer simulations, and physical systems.

Limitations. Various definitions and interdisciplinary interpretations tautologically complicate emergence as a field of study. Emergence cannot be quantified with a generalized protocol, because it is unique depending on the specific SoS being explored.

General applicability to SoS problems. Emergence is one of the traits explicitly defined in SoS literature, which instantiates it as a property of SoS.

Evolutionary Biology

Founder/Early Pioneers. Julian Huxley, 1942.

Underlying principles. Biology, automata, and emergence.

Well-known applications. Market evolution, Darwin's applications to biological beings, and ecological studies.

Limitations. Evolutionary biology not able to be validated, only retroactively verifiable.

General applicability to SoS problems. Most observable in the transformation and expansion of collaborative SoS over time.

Game Theory

Founder/Early Pioneers. John von Neumann, 1944.

Underlying principles. Probability, reward-penalty functions, etc.

Well-known applications. Conflict modeling, group dynamic models, and simulating full and no information game scenarios.

Limitations. Rationality of players is an assumption that is not always true, especially with intelligent and self-managed players/systems. Rationality also limited by information sets available.

General applicability to SoS problems. Modeling cooperative and competitive interactions of systems in an SoS.

General Systems Theory

Founder/Early Pioneers. Ludwig von Bertalanffy, 1968.

Underlying principles. Knowledge and processes across systems engineering domain.

Well-known applications. Systems biology, chemistry, dynamics, engineering, and psychology.

Limitations. Difficult to accommodate unrelated disciplines.

General applicability to SoS problems. Interdisciplinary bridge for analyses. General Systems Theory is an inherent principle in many SoS analysis methods, like agent-based modeling.

Genetic Algorithms

Founder/Early Pioneers. John Holland, 1960s.

Underlying principles. Evolutionary biology, optimization, and random events.

Well-known applications. Design optimization, economic modeling, trajectory optimization, and Monte Carlo methods.

Limitations. Computationally intensive and potential for early convergence to a local optima.

General applicability to SoS problems. Failure-resilience modeling in directed and collaborative SoS. Can model SoS optimization problems with known starting conditions but takes significant computational resources if poorly conditioned.

Group Dynamics

Founder/Early Pioneers. Kurt Lewin, 1947.

Underlying principles. Sociology, individuality, human psychology, and biopsychology.

Well-known applications. Social networks, team matching, and crowd behavior modeling.

Limitations. Conflict management is resource and time intensive.

General applicability to SoS problems. Systems within an SoS can be treated as groups or members of a group.

Information Theory

Founder/Early Pioneers. Claude E. Shannon, 1948.

Underlying principles. Communications, entropy, signal processing, and noise.

Well-known applications. Encoding and decoding algorithms, and compression methods.

Limitations. Information reliability is ignored. Signal is assumed as the only factor in transmission.

General applicability to SoS problems. Coherent with Maier's trait of geographic distribution. Explains relation between complexity and entropy.

Insect Robots

Founder/Early Pioneers. Rodney Brooks, 1994.

Underlying principles. Biology, artificial intelligence, and multi-agent systems.

Well-known applications. Environment sampling and monitoring, search and rescue, and aerial surveillance.

Limitations. Ornithopters require powerful batteries for high endurance flights which can impact weight and flight dynamics.

General applicability to SoS problems. Insect robots act as a swarm in a collaborative SoS.

Mathematical Genetics

Founder/Early Pioneers. Ronald Fisher, 1930.

Underlying principles. Evolutionary biology, genetics, and sampling.

Well-known applications. Modern synthesis, genetic algorithms, and virology.

Limitations. Requires large timescales and numerous generations. Mutations are not necessarily beneficial to genetic fitness.

General applicability to SoS problems. Genetic algorithm optimization is a direct application of this concept in network and SoS design.

Multi-Agent Systems

Founder/Early Pioneers. Reid Smith, 1980.

Underlying principles. Autonomy, control theory, system architectures, and information visibility/availability.

Well-known applications. Supply-chain management, purchasing agents, artificial intelligence programming, coordinated defense systems, and information retrieval (online search).

Limitations. Computationally expensive to implement. Complex testing, verification, and validation process. Cannot model physical systems that cannot be decomposed into components with objectives.

General applicability to SoS problems. MAS can be applied to large-scale network SoS to model interactions and protocols to observe information exchanges (e.g., internet).

Neural Networks

Founder/Early Pioneers. Warren McCulloch, 1942, and Frank Rosenblatt, 1958.

Underlying principles. Biological neural networks.

Well-known applications. Identifying patterns and trends: Sales forecasting, target marketing, risk management, process control, and biometric and recognition systems.

Limitations. Requires large data set for accurate predictions. Computationally expensive to execute in real-time. Uncertainty quantification and noise in data are big challenges.

General applicability to SoS problems. Prediction of future states of SoS based on historical data helps in designing architectural changes for better outcome (e.g., security systems, fraud detection systems).

Nonlinear Dynamics

Founder/Early Pioneers. Henri Poincaré, 1912.

Underlying principles. System and rigid body dynamics.

Well-known applications. Modeling weather, healthcare, and large transportation systems.

Limitations. Physical application is challenging. Confounding variables invalidate models easily.

General applicability to SoS problems. Differential equations can be used to model specific relationships between systems within SoS with nonlinear dynamics principles.

Object Oriented Programming

Founder/Early Pioneers. Ole-john Dahl and Kristen Nygaard, 1960s.

Underlying principles. Procedural programming, inheritance, encapsulation, and polymorphism.

Well-known applications. Implementation of Graphical User Interfaces (GUIs), web and mobile applications, and Agent-Based Modeling (ABM).

Limitations. Steep learning curve, large file structure and size, slower than procedural programming methods, and may not be applicable to all types of programs.

General applicability to SoS problems. Model and simulate independent entities and constituent systems of SoS with self-contained data, system properties, and user defined constraints (e.g., ABM).

Operations Research

Founder/Early Pioneers. A. P. Rowe, 1937.
Underlying principles. Continuous and discrete optimization, and time-variant functions.
Well-known applications. Scheduling, routing, queuing, resource allocation and inventory management in various industries.
Limitations. Operations Research (OR) takes only into account those elements and factors that are quantifiable.
General applicability to SoS problems. Design optimization of network and interactions through operations research is vital at every level of an SoS.

Political Economics of Conflict

Founder/Early Pioneers. Thomas Schelling, 1960.
Underlying principles. Economics, game theory, and cost-benefit analysis.
Well-known applications. Bargaining and negotiations, strategy planning, and political campaigning.
Limitations. The modeling of value functions of other systems/players is crucial to strategizing and bargaining, but it is under the conditions of an incomplete information environment/game.
General applicability to SoS problems. Utilized in understanding the negotiations between multiple entities and the decision-making processes to deploy those assets (e.g., multi-national policy conflicts).

Psychology

Founder/Early Pioneers. Many.
Underlying principles. Biology, behavioral sciences, and neuroscience.
Well-known applications. Weber's law, Donder's experiment, and sociology.
Limitations. Human behavior perception is complex and many psychological tools are constantly being revised.
General applicability to SoS problems. Critical to any human-centric design for any SoS that includes humans.

Self-Organized Criticality

Founder/Early Pioneers. Many.
Underlying principles. Power law distributions, and small/large event relationships.
Well-known applications. Earthquake detection/warning, electrical grids, neural nets, and economics.
Limitations. Ignores why events/outcomes happen. Timing specific events is difficult.
General applicability to SoS problems. Can be used to study system emergent behaviors.

Sociology of Teams

Founder/Early Pioneers. Michel Callon, Bruno Latour, and John Law, 1890s.
Underlying principles. Human psychology, sociology, behavioral sciences, and group dynamics.
Well-known applications. Geopolitical analysis, conflict resolution, mob violence analysis, and economics.
Limitations. Does not account for pre-existing structures, or how a network node may benefit from its location in a network.
General applicability to SoS problems. Social network analysis, characterizing relationships between nodes.

Swarm Intelligence

Founder/Early Pioneers. Gerando Beni, Suzanne Hackwood, and Jim Wang, 1988-89.
Underlying principles. Biology, group dynamics, control theory, and consensus.
Well-known applications. Computer simulations, routing and planning, autonomous surveillance, military operations, and space exploration.
Limitations. Accounting for emergent behavior is difficult in large swarm networks with independent and self-managed systems.
General applicability to SoS problems. Enables analyzing and modeling the behavior of SoS architectures where the systems are in motion and geographically spread (e.g., traffic management).

Systems Dynamics

Founder/Early Pioneers. Jay Forrester, 1961.
Underlying principles. Control theory, nonlinear dynamics, stock/flow models, and feedback models.
Well-known applications. City planning, world dynamics, and organizational dynamics.
Limitations. Assumptions generate many inaccuracies. Qualitative variables add confusion to modeling paradigm.
General applicability to SoS problems. Models geographic distribution and emergent behavior. Can model directed, collaborative, and virtual SoS.

World Models

Founder/Early Pioneers. Dennis and Donella Meadows, 1972.
Underlying principles. Simulation and analysis, and predictive methodologies.
Well-known applications. Population studies, environmental predictions, and agriculture yield predictions.
Limitations. Ignores potential changes to capital markets. Labor contribution is misrepresented.

General applicability to SoS problems. Exhibits operational/managerial independence, evolutionary development, and emergent behavior. Acknowledged SoS objectives, funding, authority, and management in parallel.

In the rest of Part II of this text, following this chapter, we dive deep into a select number of modeling and analysis methods that are most central to the widest class of SoS problems. For modeling, we present relevant concepts from the field of applied network science (Chapter 5) and agent-based modeling (ABM, Chapter 6). Though by now it should be obvious to readers, the connectivity and interaction of SoS (features well represented in network science) and the independence of SoS participants (exactly the essence of ABM) are appropriate foundations. For analysis, we present in detail a family of analytic methods and tools designed especially for SoS problems (Chapter 7) and couched in relation to each other in what became known as the "SoS Analytic Workbench." In Chapter 8, we examine how recent advances in emerging analytics can and do play an instrumental role in SoS M&A and problem solving.

Before turning to next chapters, we close this chapter with a section that offers our advice for projects, providing learners of SoS M&A, whether students, researchers, or practitioners, an opportunity to "put it all together", by applying the DAI process on an SoS problem from start to finish.

4.3 PUTTING IT ALL TOGETHER: THE SOS M&A PROJECT

The semester-long project has been an anchor of the AAE 560 course since its inception. As a team project, it brings both great learning opportunities and the typical challenges associated with team work. This is appropriate, since SoS M&A, and SoSE at large, is a "team sport" just as SoS itself is a collection of collaborating components. Since 2012, the AAE 560 course has been included both on-campus students as well as distance-learning students. In recent years, there has been healthy enrollment in both categories, averaging 80 on-campus students and 30 distance students. While the lecture delivery to distance students is asynchronous (not unusual since these are working professionals and thus not typically able to attend lecture during normal business/lecture hours), the active use of instructor curated discussion websites and modern collaboration/communication tools makes teams consisting of on-campus and distance students no more challenged than homogeneous teams. In fact, the presence of practicing professionals in the course enhances the learning of the on-campus students (and the instructors). The insight and practical experiences these professionals bring to the many on-campus learners who generally have had less 'real-world' experience is of unique value. The project experience is where this mutual-learning among course collaborators flourishes the most.

The following sections present the documents provided to students, to guide them through the project experience, and establish the learning objectives and expected deliverable items. Instructors can follow these documents to provide appropriate feedback to students involved in SoS projects.

4.3.1 OVERALL PROJECT DESCRIPTION

Working in teams of 2-4 people, the team project allows learners to demonstrate knowledge of key concepts through application, culminating in a final report usually expected in the form of draft peer-reviewed article as well as en engaging poster/presentation delivered to the entire class. Participants self-select into teams according to their interest in project topics provided by the instructor (or self-generated, in consultation with the instructor). Teams must successfully apply the DAI process.

4.3.1.1 Overview

While the project typically culminates in two major end products, an "interactive poster" and a final report cast in the form of a draft peer-reviewed journal article, the interim products from each DAI phase pave the pathway to this destination. Students self-select into teams and embark on application of the DAI approach. Teams launch their project based upon a paper, report, etc. that addresses an SoS challenge in their area of interest.

4.3.1.2 Project Learning Objectives

The learning objectives associated with a project are to:

- Get hands-on (or "collective brains-on") experience with problems of the SoS-type.
- Formulate System of Systems design problems.
- Develop/use appropriate M&A methods for these problems.
- Learn in detail at least one method for SoS M&A taught in the course or found in the universe of system modeling approaches.
- Gain experience with working in teams on SoS challenges.

4.3.1.3 Project Technical Objective

Starting from the selected work, teams use the DAI process to first assess the scope/quality/validity of the SoS modeling approach and then define an "extension" of the problem/approach that becomes the core objective of the AAE 560 project. This extension could take the form of modifying the SoS problem and then adapting the presented method to address this modified problem. Alternately, a team could keep the selected work's core problem, and use DAI to build (and verify and validate) a different kind of model. In any case, teams must probe in depth one or more of Maier's four heuristics for SoS architecting by exploring a design space consisting of design variables from among the three classes of SoS design variables. Since the selected work, most likely, will not have leveraged Maier's heuristics and these three classes of variables, teams must be thorough in translating them to your selected work's context.

4.3.1.4 Project Deliverable Items

The three deliverable items correspond to the 3-phase DAI process.

Deliverable 1: Definition Phase. Use the definition phase methods and tools to critique the definition of the SoS challenge/problem in the selected work and produce a superior one to describe their problem (or the modified one you wish to tackle). Key elements of this deliverable include not only a specification of the problem space (e.g., ROPE Scope, other definition phase items, etc.), but also a specific research question related to a specific problem from the chosen domain.

Deliverable 2: Abstraction Phase. In the abstraction phase, teams use a variety of techniques to represent alternate architectures for the SoS challenge. Targeting support for the implementation phase, key elements of this deliverable include abstraction of component systems, their organization/interdependence, and degrees of control as well as the specific hypotheses that guide the specific modeling approach to be implemented.

Deliverable 3: Implementation Phase. The final phase of the project builds a model and conducts analysis to test hypotheses and thus answer the SoS research questions. It typically culminates in the two end product items: (a) an interactive poster that will be shared with the class and reviewed by fellow students, (b) a final report in form of a draft journal article. Each of these two end products must address the execution of the whole DAI process: what the original selected work offered, how the problem/hypothesis was modified, why it's important, the model and analysis built, evidence and findings, and conclusion.

4.3.2 DEFINITION PHASE DELIVERABLE

Use the definition phase methods and tools to critique the definition of the SoS challenge/problem in your selected work and produce a superior one to describe their problem or the modified problem you wish to tackle. Key elements of this deliverable include not only a specification of the problem space (e.g., ROPE table, other definition phase items, etc.), but also at least two candidate research questions related to a specific problem from the chosen domain. Remember Bellman's imperative: "the right problem is always so much harder than a good solution."

4.3.2.1 Format

A four-page report that satisfies the deliverable description above and contains and integrates the key elements outlined below. (Format: single-spaced, 12-point font. Bibliography is excluded from page limit.)

4.3.2.2 Procedure

Read, Learn, Discuss, Scope. These are the keys for executing this phase. Probe deeply your selected work, and then related work, to solidify your understanding of the SoS problem and express it clearly in our SoS lexicon. Your report must contain these, and only these, three sections; use the exact words below as section titles (e.g., every team's report will start with "A. Operational Context"; you can use whatever sub-sectioning). Your text must logically relate each section to the next, not separate disconnected products.

1. Operational Context: Describe the aspects (e.g., modes, time scales) of your selected SoS(s) in terms of how it operates today (or, if it does not exist yet, what are the key goals for it). Who are the relevant stakeholders and how do they influence operation, etc.? What important items were missed by the authors?
2. Status Quo and Barriers: What is the current state of the SoS(s) that are part of your problem domain and what are some perceived barriers and limitations of it that would motivate a better SoS architecture (e.g., it is fragile, inefficient, not scalable, has unknown dynamics, is prone to negative emergent behavior, etc.)? If the SoS does not yet exist, what are the known barriers to its realization?
3. Problem Definition (Scope Categories and Levels): (a) Create and describe a ROPE table: Translate the selected work's problem definition into our SoS lexicon (e.g., a ROPE table, associated design variable classes, etc). Bear in mind that your usage of lexicon should be clear and concise. Build some details that describe (in a separate place) each box/element in the ROPE table, tagging references or interesting thoughts that have arisen within your team. (b) Generate two specific, valid research questions related to your selected work; these may be ones already provided or they may be new ones derived from the content of your selected work. Highlight which portion of the ROPE table is relevant to each RQ.

4.3.3 ABSTRACTION PHASE DELIVERABLE

In the abstraction phase, teams use a variety of techniques to represent alternate architectures for the SoS challenge. Targeting support for the implementation phase, key elements of this deliverable include abstraction of component systems, their organization/interdependence, and degrees of control as well as the specific hypotheses that guide the specific modeling approach to be implemented.

4.3.3.1 Format

Prepare a four-page report that satisfies the deliverable description above and contains and integrates the key elements outlined below. (Format: single-spaced, 12-point font. Bibliography is excluded from page limit.)

4.3.3.2 Procedure

Focus on clarifying SoS hierarchy and interaction descriptions. These are the keys for executing this phase. Probe deeply your definition phase products and your original selected work to proceed to a successful abstraction phase, culminating in hypotheses. Your report must contain these, and only these, four sections. Use the exact words below as section titles. Your text must logically relate each section to the next, not separate disconnected products.

1. Abstraction Representation: Present and discuss your abstraction depiction, i.e., a kind of "paper model." Make sure that the most important ROPE table elements are evident along with some indication of their key behaviors/interests/constraints that drive the SoS dynamics as well as drivers, disruptors, and, perhaps most importantly, their interdependencies in networks.
2. SoS Design Variables (DVs): Considering your Abstraction Representation and your Research Questions, describe your most relevant SoS DVs and their interrelation across the classes of Composition, Configuration, Control. In other words, characterize systems that may compose the SoS, their possible configuration with interdependencies expressed in architecture, and the key variables that may distinguish alternate SoS configurations in terms of independence/control in your setting.
3. Candidate Performance Metrics and Hypotheses: Develop a list of candidate metrics to evaluate alternatives and that will be manifested as measurable quantities in your computer model/simulation. Draft preliminary hypotheses (building from your RQ), using these metrics for your SoS design that can be tested in a model.
4. Modeling Approach: Specify the modeling and analysis approach you intend to take. Bear in mind two considerations: (1) your approach may involve a hybrid or combination of approaches discussed in class, on Solberg Chart, or from elsewhere, (2) there will be a chance to modify the modeling approach after digesting all the options and the pros/cons.

4.3.4 IMPLEMENTATION PHASE DELIVERABLE

The final phase of the project builds a model and conducts analysis to test hypotheses and thereby answers the SoS research question. It typically culminates in two deliverable items: (a) a poster suitable for interactive discussion that will be shared with the class and reviewed by fellow students, (b) a final report in the form of a draft journal article. Each of these two must address the execution of the whole DAI process, with the poster especially highlighting: What was the original selected work used to start your project? How the problem/research question/hypothesis from that work was modified, expanded by you? Why it is an important SoS problem? What model did you build and what analysis did you conduct? What evidence supports your findings (including some verification and validation)? What are the conclusions?

Simulation/analysis with your model must properly characterize the salient SoS design variables inherent in your hypotheses and include the emergent properties that arise when individual systems exhibit operational and/or managerial independence. You should highlight one or more of Maier's four heuristics for SoS architecting that are part of your implementation product. In other words, this is not just another model/study you built, but it is specific to SoS M&A.

4.3.4.1 Deliverable Items and Expectations

General: Your implementation phase deliverable items must be comprehensive, covering the main products you developed in the first two phases as well as answering:

1. How well did your baseline SoS architecture perform with respect to meeting the initial capability requirements of your SoS problem? Use the metrics you defined in the abstraction phase to answer this question.
2. What evidence did you produce using your analysis that resulted in accepting or refuting your hypothesis? This should include a summary of the recommended SoS architecture uncovered by your analysis.

Specific: Following are important specific items to address in both the final report and poster presentation:

1. Briefly describe again the elements you identified in the definition phase: operational context, status quo, barriers, ROPE scope.
2. Use lexicon and taxonomy learnt in class to summarize your problem. Also state your main research question(s).
3. Briefly describe/depict the paper model/abstraction of your project, and the main hypotheses.
4. Specify the modeling approach you decided to take and justify your selection.
5. Describe your work done in the implementation phase, including model V&V, clear answers to hypotheses posed, a reflection on results, and answers to the other questions asked in this document.

4.3.4.2 Report Specifics

Prepare your report in a format applicable to a journal/conference that best suits your work. This will encourage you to think rigorously about your writing and, perhaps, some may have something worth publishing. State the formatting standard you choose as a note with your submission. Whichever format you select, the report length should not exceed 10 pages of single-spaced, 11-pt. font size, including tables and figures (put figures and tables in-line with text). You can also include an appendix separately with "the rest of the story"... no more than 5 pages for that. References are required but not included in this length limit. All of the main content requested (items above in the *Specific* section) must be contained adequately in the main 10 pages.

4.3.4.3 Poster Presentation Specifics

Surely all students have attended and likely presented at a "poster session." Use common sense good practices, for example use a good sized template for a poster (e.g., if printed, it would be like 2x3 feet). The poster should clearly but comprehensively address the items in the *Specific* section above. It must be understandable by someone who is reading it without hearing the 'presenting' of it. Teams may include a recording of the presenting of the poster; that is frequently well-received, but it is not a requirement.

In preparing poster presentations, consider a more "brass tacks" approach, especially to form good questions and present actionable outcomes. One of the most famous and succinct approaches is the "Heilmeier Criteria" promulgated by former

DARPA Director George Heilmeier as part of his "Catechism" for DARPA program managers and partners [94].

4.4 NOTES ON APPLICATIONS AND SELECTED LIST OF SOS M&A PROJECT TOPICS

Table 4.2 is a small sample of topics that teams of AAE 560 SoS M&A students have undertaken over the years organized into topical areas (themselves only a small sample of the diverse domains in which projects were sourced). An asterisk (*) indicates those "historic" projects/teams that executed work in the first ever offering of the course in 2006. One memorable project from that inaugural year built a web crawler to gather data from a then-little-known social network company called Facebook (now Meta). They used network analysis to expose the nature of connections among students on the Purdue campus. None of us imagined what this kind of platform, and analysis, would evolve into. Also, four projects from this 2006 group appear in italics indicating that they turned into master's thesis or PhD dissertation topics. This happy outcome would repeat itself many times in the ensuing years, a very gratifying outgrowth and a sign of success in creating "SoS thinkers, modelers and doers" of the future.

Two additional observations about this list are in order. First, while only a handful of projects are listed under Integrated Defense, this area was the subject of many projects over the years (third in number only to transportation and space exploration). Perhaps, this reflects the fact that Integrated Defense has been, arguably, a most active domain for development, research, and advancement of SoS innovation and SoSE application. In fact, it is an area where a large component of our own research work has occurred, from SoS information fusion in ballistic missile tracking [142], to formulations optimizing networks of independent sensors, weapons for variable targets [58], and architectures for consensus and orchestration [123, 134]. Yet, in Part III of this book, we chose to give deep application treatment in two civil/commercial areas: Air Transportation and Space Exploration Systems. Our reasons are two-fold: (a) given the ubiquity of integrated defense-inspired research and application, there are ample materials (papers, reports, etc.) for interested learners to consult, (b) air and space transportation are both more easily accessible by large swaths of practitioners as well as SoS-interested research and education communities. We also feel that these two application areas are well-suited to the "aspirational" goals inherent in the "SoS way of thinking."

Second, for the reader interested in individual sources of SoS applications, there are a small number of entire volumes dedicated to SoS applications. Once again, two of the most important of these were edited by SoS pioneer Professor Mo Jamshidi. First published in the same year (2009), these two edited volumes on System of Systems Engineering [107, 108] remain important sources for diverse coverage of SoS methodology, and especially case study/application experiences. We still recommend that newcomers to the field make good use of these two resources. Per our point in the previous paragraph, these two volumes are also well-populated with integrated defense (and related) examples.

Table 4.2

Small sample of the diverse set of AAE 560 *System of Systems Modeling & Analysis* team project titles. Italic font identifies projects that developed into master's theses or PhD dissertation topics. Asterisks identify projects and teams from the first offering of the course at Purdue University in 2006

Transportation
*Multi-modal Network Design, Modeling and Optimization**
Impact of Automatic Dependent Surveillance Broadcast (ADSB) on Air Transport Networks*
Urban Air Mobility
Urban Transportation System-of-Systems: Implementation of Congestion Reduction SoS
Safety analysis for early investment in Intelligent Transportation System (ITS)
Space Exploration
Mars Human Exploration Resource Supply Chain*
*Solar System Mobility Network**
Asteroid Mining Opportunities Analysis
Transportation Infrastructure of Lunar Base Exploration
Orbital Refueling System
Migration of International Space Station: SE to SoSE
Food and Water Management Impact on Space Exploration
Supply Chain/Manufacturing/Logistics Management
Commercial Aerospace Supply Chain System of Systems Horizontal vs. Vertical Integration
Systems of Systems Analysis on Milestone Performance in NASA Program Lifecycles
Modeling and Analysis of Small Network IoT Security
Comprehensive Testing Method for Engine Performance Analysis
Emergency and Disaster Response
*Disaster Relief with Regard to Critical Infrastructure**
Emergency Wildfire Response SoS
Manned-Unmanned Teaming (MUMT) for Undersea Search
Resilient Renewable Energy SoS
Healthcare/Food
*Health Informatics**
More Efficient Food Distribution Network
Ensuring Collaboration in a Hospital Urgent Care SoS
Global Food Security in 50 Years
Integrated Defense
Optimal Network Topology for Defending a Target*
U.S. Coast Guard Deepwater/SAR Acquisition*
Rethinking Ballistic Missile Defense System
Air Force Space Command Missile Warning System of Systems
Navy Fighter Jet Fleet Transition via SoS
Other
Purdue Social Network Mapping*
Better Approach to Riot Response by Police via SoS

4.5 CHAPTER SUMMARY

Good modeling practices and good research approaches go hand in hand in SoS M&A because of the increased risk of "scope creep" starting from the very first step of problem definition. In this chapter, we refreshed these fundamental good practices, offered insights from our experience in communicating them to new learners, and provided examples of their use from prior AAE 560 course projects. We also presented the briefest of information capsules on each of the methods covered in the "Solberg Chart" (Figure 1.1) as a possible template research teams could use when deliberating on selection of an implementation phase method. Finally, the four documents that define the semester-long project in AAE 560 course are presented. Included is an overall project guide and a specific guide and deliverable item definition document for each of the three DAI phases. Beyond instructors and students, these documents and the structured approach they entail are useful for practitioners embarking on an SoS M&A activity.

5 Network Theory

The 3-phase DAI process of SoSE begins with the definition phase during which we identify the resources which constitute the SoS. Along with the resources, we also identify collections of resources organized as networks at higher levels of the ROPE framework. During the following phase of abstraction, we further refine our representation of an SoS as a network to identify and represent the interactions among its resources. Indeed, the pursuit to identify and design an SoS' emergent behaviors leads us to architect the structure of system interactions. This representation of an SoS as a network, with constituent systems as nodes and their interactions as links, comes from network theory, which offers a versatile and capable model for both the representation and analysis of a system of interacting systems.

Most complex systems can be abstracted and represented as a network [31], whereupon we can set up the problem of SoS analysis to be evaluated using the vast theory of network science. For example, knowledge of network topology enables insights on implications of set of linkages on the system's characteristics such as robustness to failure or trade-off between efficiency and throughput. Network growth algorithms may help predict future connections within the system. Ultimately, using network theory can help us with SoS analysis provided we identify appropriate mappings between a network's topological characteristics and its operational performance metrics, and use the right dynamic algorithms to predict future evolutionary changes to the topology.

In this chapter, we introduce the basic concepts of network theory including the properties used to describe networks and some applications of network theory to study real-world systems. In particular, we cover material that is relevant to the use of network theory in SoS modeling and analysis. The study of networks is a wide field and this chapter forgoes depth in favor of the breadth of material; for more detailed treatment of networks, refer to [10, 136]. Our primary objective in studying this chapter is to learn how network theory can be applied to modeling and analysis of SoS.

5.1 BASIC GRAPH THEORY AND NETWORK MEASURES

Network theory finds its origins in graph theory, a part of mathematics which originated, famously, from the problem of the Seven Bridges of Königsberg and has since grown tremendously [15, 113]. Mathematically, a graph $\mathscr{G}$ is comprised of a set of vertices, $\mathscr{V}$, and a set of edges, $\mathscr{E}$, with each edge connecting a pair of vertices. It is versatile enough to be used to represent a wide variety of real-world systems and their interconnections. For example, vertices may represent a person, an aircraft, or a webpage. Correspondingly, edges may represent friendship between two persons, a route over which an aircraft flies, or a hyperlink connecting one webpage to another.

In our discussion hereafter, we will use the terms network, nodes, and links to refer to a graph, vertices, and edges, respectively, though we will continue to use the above symbols to represent them.

Let us begin with a most general representation of a real-world system in which we strip away all details of its components and their interactions and assume that they are alike. For example, suppose we want to identify the social network among a group of individuals without being concerned with who they are or the type of relationship that connects them. To do so, we can abstract away all details of the specific individuals and represent them using a common denotation as nodes, and add a link between two nodes if the individuals they represent have any type of relationship. This kind of abstraction is widely used in social sciences and can be used to map the network of relationships among all individuals in our set and analyze them using any of the measures we will discuss later.

While we can analyze various graph theoretic attributes of a network by abstracting away domain-specific details of nodes and links, network analysis becomes much more useful when we assign characteristics to the nodes and links corresponding to the attributes of the systems that we want to represent. Variation of the type of nodes or links is one source of classification of networks and, hence, the corresponding systems. In turn, classifying networks on the basis of differences between types of nodes and links will allow us to choose appropriate network models and measures for the systems that we wish to represent. The following subsection discusses some of the classes of networks based on the types of their nodes and links.

5.1.1 TYPES OF NETWORKS

5.1.1.1 Undirected Network

Without being concerned with the characteristics of an individual, let us assume that if person A is a friend of person B, then person B is also a friend of person A. That is, friendship among two individuals is bi-directional.

An *undirected network* is one in which the links lack any direction, which means that all links can be traversed in both directions. That is, a network $\mathscr{G}$ is undirected if for any pair of nodes $(u,v) \in \mathscr{V}$, we have $u \rightarrow v \Leftrightarrow v \rightarrow u$, where the symbol "$\rightarrow$" indicates a link exists with its direction indicated by the arrowhead.

Figure 5.1(a) shows an undirected network. Examples of undirected networks include the network of actors who have worked together in a movie, the electric grid, because electricity can be made to flow in either direction of a link, and the air transportation network, because flight routes can be traversed in either direction.

5.1.1.2 Directed Network

Suppose we want to map the set of phone calls made among friends over a period of a day. In such a case, we can represent the individuals as nodes and the phone calls as links with directions indicating who called whom.

A *directed network* is one in which links have a direction meaning that links can only be traversed in the direction indicated. That is, if a network $\mathscr{G}$ directed, it means

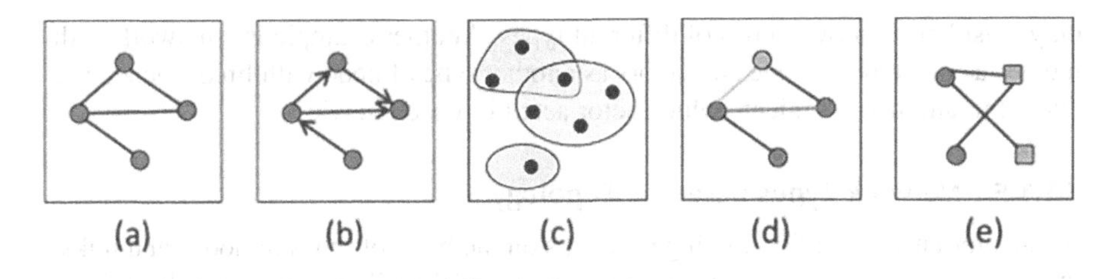

Figure 5.1: Types of graph: (a) undirected, (b) directed, (c) hypergraph, (d) heterogeneous, and (e) bipartite.

that there exist some pairs of nodes $(u,v) \in \mathscr{V}$, for which we have $u \to v \nLeftrightarrow v \to u$. However, not all links in a directed network need to be uni-directional, some of the links may be bi-directional while the others may be uni-directional.

Examples of directed networks include the World Wide Web, because the hyperlinks point from one web page to another and thus have a direction, and the set of flights recorded over a given time period. Figure 5.1(b) shows a directed network.

5.1.1.3 Hypergraphs

In both directed and undirected networks, links connect a pair of nodes. However, we may define links which connect more than two nodes – such links are called hyperedges. *Hypergraphs* are a type of network in which the links may connect more than two nodes.

For example, from among a set of individuals, we can identify groups of friends rather than just pairs of friends using hypergraphs. Similarly, in an organization, teams of individuals can be connected with hyperedges, where each hyperedge connects all members of a team. A node, which is an individual employee, may be part of multiple hyperedges. Figure 5.1(c) shows an example of a hypergraph.

5.1.1.4 Bipartite Graphs

So far, we assumed that all nodes in a network are alike and the network types were based on differences among links. This is not true in real networks. *Heterogeneous* graphs are those in which there are different types of nodes (Figure 5.1(d)). For example, in an organizational network where nodes represent employees, we can differentiate among nodes by the type of position of employees. In a technological network such as the internet, we can distinguish nodes as the type of device which connects to the network.

A special class of such heterogeneous graphs are those classified as *bipartite*, in which there are only two different types of nodes, with no links between two nodes of the same type (Figure 5.1(e)). Network $\mathscr{G}$ is *bipartite* if nodes V are to be partitioned to two subsets M and N such that each link of G connects a node of M to node of N.

For example, we may want to identify the friendships among men and women. In this case, we can generate a graph with two different types of nodes where links only exist between two nodes of different types. Another example is a network with movies as one type of node and actors as another type of node with links connecting one set to another to indicate which actor acted in which movie.

5.1.1.5 Network Types Based on Topology

As an alternative to differentiating networks on the basis of types of nodes and links, we can classify networks based on the topology of their links. For example, a *fully-connected* network is one in which every node is connected to every other node. In case of directed networks, we refer to them as strongly connected; i.e., a directed network is strongly connected if for every pair of nodes (u, v), there is a path $u \rightarrow v$ and $v \rightarrow u$.

A simple and commonly studied type of network is called a *random network*, which is a network in which nodes connect with each other with a fixed probability p. In such a network, the degrees of nodes are distributed according to a binomial distribution; we will discuss node degree in the next section. Random networks were first studied by Erdős and Rényi and formed the basis of the random graph theory which was a new field of mathematics when it was first proposed. An advantage of random networks is that they are relatively easy to model mathematically. We will discuss formation of random network models in the section on network growth algorithms.

Real-world networks are mostly found to be very dissimilar to the random graph in their degree distributions. In fact, we can exploit the non-random behaviors of real networks so as to reveal the possible mechanisms that lead to the formation of their structure and their behaviors. A more common type of network topology found in the real world is called a *scale-free network* in which nodes connect preferentially to other nodes of higher degree. The degrees of nodes of a scale-free network follow a power law distribution. Figure 5.2 shows examples of fully-connected, random, and scale-free networks.

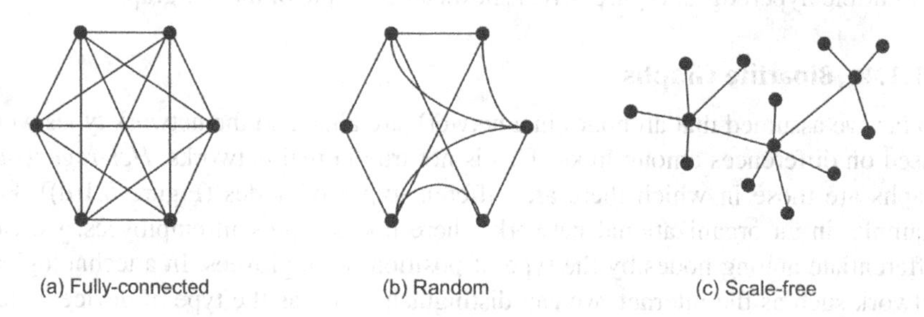

Figure 5.2: Types of graphs based on topology (from left to right): fully connected, random, and scale-free.

5.1.2 MEASURES IN NETWORK THEORY

Disregarding the details of node attributes, in this subsection, we will look at some of the metrics used for measuring attributes of a network's topology. All of these metrics impact their behaviors which will be helpful for our analyses.

Figure 5.3 shows two versions of an example network which we will use to illustrate the calculation of metrics throughout this section. On the left is a five-node network with undirected links, while on the right is the same network but with directed links. Both of these networks have $N = 5$ nodes. The undirected network has $L = 6$ links, whereas the directed network has $L = 10$ links, because we count links pointing in different directions separately.

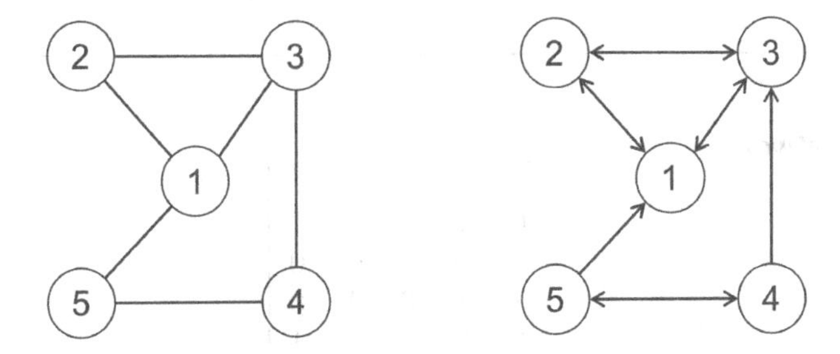

Figure 5.3: Example network for calculation of network metrics.

Smaller networks such as those in Figure 5.3 can be easily visualized and we can observe the set of sub-system connections. We may even be able to derive qualitative insights based on such visualization. However, real-world networks can have millions of nodes and links, in which case graphical visualization, while possible, would not be helpful in analysis. For large networks, we can use the tools and theory of network science to analyze both attributes of network topology and their dynamic behaviors. For example, the connections among systems affect the network's attributes such as robustness, resilience to failure, cost of operation, etc. With network growth algorithms, we can evaluate the dynamic behaviors of a network, such as the evolution of its topology over time as a result of addition or removal of links.

First, we need a compact way to represent a network. We do so by using an adjacency matrix, A, which is a matrix with N rows and N columns where N is the number of nodes in the network. Entries of this matrix are $a_{i,j} = 1$, if there is a link from j to i, and $a_{i,j} = 0$, otherwise.

The adjacency matrices of the two networks in Figure 5.3 are shown below. Table 5.1 shows the adjacency matrix for the undirected network, while Table 5.2 is for the directed network.

Notice that in case of the undirected network, the adjacency matrix is symmetric, which is not necessarily true for directed networks. Further, our example network is unweighted; in case of a weighted network, the entries in the adjacency matrix can indicate link weights. For example, $a_{i,j} = w_{i,j}$ will indicate the weight of the link from node j to node i.

Table 5.1
Undirected Network

	1	2	3	4	5
1	0	1	1	0	1
2	1	0	1	0	0
3	1	1	0	1	0
4	0	0	1	0	1
5	1	0	0	1	0

Table 5.2
Directed Network

	1	2	3	4	5
1	0	1	1	0	1
2	1	0	1	0	0
3	1	1	0	1	0
4	0	0	0	0	1
5	0	0	0	1	0

From the adjacency matrices we can verify the total count of links in each network as follows. For the undirected network, we sum entries in the upper triangular or the lower triangular matrix to avoid counting the same link twice. Thus, the total number of links is

$$L = \sum_{i<j} A_{i,j} \tag{5.1}$$

which for our example network totals 6, as expected. For the directed network, we sum all entries in the adjacency matrix

$$L = \sum_{i,j} A_{i,j} \tag{5.2}$$

which equals to 10.

5.1.2.1 Degree, Degree Distribution, and Network Density

We start with a basic measure that is a node's degree and the network's degree distribution.

The *degree* of a node is the number of links connecting that node with its neighbors. In the case of an undirected network, this is simply a count of the total number of links connected to the node. For directed networks, a node's *in-degree* and *out-degree* can be calculated separately as the number of links pointing toward and away from the node, respectively. The degree of a node is a measure of how important a node is to the network. Large nodes, the ones with a high degree relative to other nodes in the network, are called *hubs*. Hubs, as we will discuss shortly, are important to a number of network phenomena such as spreading of information within the network or the robustness of a network to failures.

Let us represent the degree of a node i as k_i. The average degree of a network is the mean of degrees of all nodes in the network; similar to node degrees, the average degree of a directed network can be calculated separately for in-degrees and out-degrees of its nodes. The average degree of an undirected network is:

$$\bar{k} = \frac{\sum_i k_i}{N} = \frac{2L}{N} \tag{5.3}$$

where N is the total number of nodes and L is the total number of links in the network. The factor of 2 in the numerator accounts for the fact that each link is counted twice in an undirected network when summing degrees of each of its nodes. The equation for average degree of a directed network drops the factor of 2 in the numerator, so that, for a directed network:

$$\bar{k} = \frac{\sum_i k_i^{in}}{N} = \frac{\sum_i k_i^{out}}{N} = \frac{L}{N} \tag{5.4}$$

Useful insights can be gained by plotting the degree distribution of a network. We can do so by tabulating the data on degree of each of the nodes in the network, then normalizing the degree distribution as $p_k = N_k/N$, where p_k is the fraction of nodes with degree k and N_k is the number of nodes of degree k, and finally plotting its histogram; see Subsection 5.1.2.1.1. We can interpret p_k as the probability that a randomly selected node has degree k.

5.1.2.1.1 Example 1

Tabulate the degree of each node in both the directed and undirected networks shown in Figure 5.3. For a simple example such as this one, we can directly count the degree of each node from the network's graphical networks. Alternatively, we can use the adjacency matrix to determine a node's degree. For the undirected network, the degree of a node is the sum of either the corresponding row or column in the adjacency matrix. For a directed network, the column sums give the nodes' out-degrees, while the row sums give their in-degrees. The following table shows the degree network for this example.

From these tables, we can plot the histograms of degree distribution as shown in Figure 5.4.

Network density is the ratio of number of links to the number of maximum possible links in the network. For an undirected network, density is given by:

$$\rho = \frac{2L}{N(N-1)} \tag{5.5}$$

Table 5.3
Degrees of Nodes of Network in Figure 5.3

Node	Undirected	In-degree	Out-degree
1	3	3	2
2	2	2	2
3	3	3	2
4	2	1	2
5	2	1	2

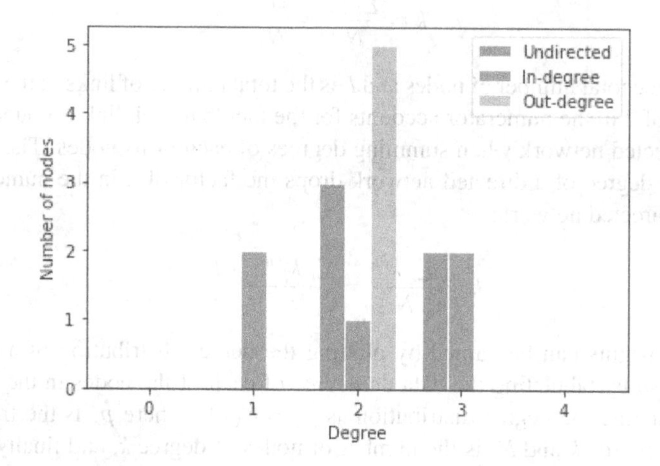

Figure 5.4: Degree distribution for the networks in Figure 5.3.

where $0 < \rho \leq 1$. The highest density of $\rho = 1$ corresponds to a fully connected network. Similar to average degree, we drop the factor of 2 when calculating density of a directed network.

5.1.2.2 Paths, Cycles, and Network Diameter

In many instances of network problems, we are concerned with flow of mass, energy, or information over the links. For example, we may want to send a message from one node to another in a communication network, or a passenger may want to fly from one airport to another over the airline network. The ordered sequence of links traversed during the course of such flow from origin to destination is a path. *Paths* are elementary walk-sequences of nodes from an initial node u to final node v where the nodes do not repeat.

Suppose, for example, we want to send a message from node 1 to node 4 in the undirected version of Figure 5.3. The message can follow the path $1 \rightarrow 5 \rightarrow 4$ or $1 \rightarrow 3 \rightarrow 4$. While we can traverse undirected links in either direction, for a directed graph, we have to follow link directions. Thus, in the directed network of Figure 5.3, the message can take the path $1 \rightarrow 3 \rightarrow 4$, but it cannot go through node 5 because the link connecting nodes 1 and 5 points in the wrong direction. A *cycle* is a path where the first and last nodes are the same. In Figure 5.3, the sequence $1 \rightarrow 3 \rightarrow 2 \rightarrow 1$ is a cycle.

A common network problem is to find a path from an origin node to a destination node within the network. For such a problem, graph search algorithms such as breadth-first search or depth-first search can be employed to find a sequence of nodes from the origin to the destination while respecting link directions. However, merely knowing the path may not be enough. A natural next task would be to calculate the cost of traversing a path. Knowing the cost would help us choose the one with a lower cost when multiple paths are found.

For an unweighted network, the length of a path is merely a count of number of links that make up the path. For weighted networks, the path length is the sum of weights of consecutive links in the path. Thus, path length is calculated as:

$$c_p = \sum_{i=1}^{l} w_{i,i-1} \tag{5.6}$$

where $w_{i,i-1}$ is the weight of link connecting node *i-1* to node i and there are l links in the path. In unweighted networks, $w = 1$.

The *shortest path* is the path that connects any two pairs of nodes in the network in the minimum number of "hops" or links in case of an undirected network, or the least distance measured as sum of weights of the links in the path. The shortest path measured in terms of number of links is also called *geodesic shortest path*.

A simple way to measure the size of a network is to count the total number of nodes. Yet another measure of the size of a network is its *diameter*, which is the length of the longest shortest path. We can also measure an average of shortest paths in the network which would indicate the connectivity in the network, with a lower value indicating higher connectivity.

5.1.2.2.1 *Example 2*

Refer again to the network shown in Figure 5.3. In the undirected network, the length of the path from node 2 to node 4 is 2 when following the path $2 \rightarrow 3 \rightarrow 4$. This is the shortest path between these two nodes. There are two other longer paths both of length 3, viz., $2 \rightarrow 1 \rightarrow 3 \rightarrow 4$ and $2 \rightarrow 1 \rightarrow 5 \rightarrow 4$. Since this is a small example, we can list all paths between all pairs of nodes and select the shortest path for each pair. You can check that the diameter of the undirected network is 2, which is also the diameter for the directed network.

5.1.2.3 Clustering Coefficient

The *clustering coefficient* of a node i, CC_i, measures the degree of connectivity among the neighbors of a give node. In other words, the higher the CC_i of a node i, the more interconnected are its neighbors.

In a network with N nodes, the clustering coefficient of node i is calculated as

$$CC_i = \frac{2L_i}{k_i(k_i-1)} \tag{5.7}$$

where L_i is the number of links among the neighbors of node i and k_i is the degree of node i. The network's average clustering coefficient can then be calculated as

$$\overline{CC} = \frac{\sum_i CC_i}{N} \tag{5.8}$$

We can also calculate a network's *global* clustering coefficient as the ratio between the number of actual triangles and the possible number of triangles, i.e., the triples centered at a node.

$$GCC = \frac{3 \times \text{number of triangles in whole network}}{\text{number of triples in whole network}} \tag{5.9}$$

In terms of network topology, transitivity means the presence of a heightened number of triangles in the network – sets of three vertices each of which is connected to each of the others. Higher transitivity means higher global clustering coefficient which indicates higher connectivity among the nodes of a network.

5.1.2.3.1 Example 3

In Figure 5.3, for the undirected version of the network, node 1 has three neighbors. The number of links among these neighbors is one – nodes 2 and 3 are connected by a link, but neither of them has a direct link to node 5. The maximum number of possible links between three nodes when we disregard link direction is three. Therefore, the clustering coefficient for node 1 is $\frac{1}{3}$.

5.1.2.4 Node Centrality

We can measure the importance of a node to its local neighborhood using node centrality measures; we do so in two ways.

Eigenvector centrality is measure of nodal importance based on principle that connections to stronger nodes are more important than connections to weaker nodes. As an example, Google's page-ranking algorithm applies this metric. This measure is calculated as,

$$x_i = \lambda^{-1} \sum_j a_{ij} x_j \tag{5.10}$$

where $A = a_{ij}$ is the network's adjacency matrix and λ is the largest eigenvalue of the adjacency matrix.

Betweenness centrality is measure of nodal importance calculated by counting the fraction of shortest path that goes through node i; this is calculated as,

$$B_i = \frac{2}{(n-1)(n-2)} \sum_{s \neq t \neq i} \frac{\sigma_{st}(i)}{\sigma_{st}} \tag{5.11}$$

where σ_{st} is the total number of shortest paths between nodes s and t, and $\sigma_{st}(i)$ is the number of shortest paths between these nodes that go through node i. In other words, the betweenness centrality of a vertex i is the number of geodesic paths between other vertices that run through i.

5.1.2.5 Assortativity

Our final metric is called *assortativity* or *degree correlation*. We can calculate four different degree correlations for each node in a directed graph. The *in-in* degree correlation is the calculation of correlation of the in-degree of the node under consideration with the in-degrees of its neighbors. The *in-out* degree correlation is for the in-degree correlation of the node under consideration with the out-degrees of its neighbors. Similarly, *out-in* and *out-out* degree correlations are for the out-degree of node under consideration with the *in* and *out* degrees of its neighbors, respectively.

If the node under consideration is indicated by subscript i, and its neighbors are indicated by subscript j, the degree correlations are given by:

$$k_{nn}^{in-dir}(k_i^{in}) = \frac{\sum_j a_{j,i} k_j^{dir}}{k_i^{in}} \tag{5.12}$$

$$k_{nn}^{out-dir}(k_i^{out}) = \frac{\sum_j a_{i,j} k_j^{dir}}{k_i^{out}} \tag{5.13}$$

Here, *dir* indicates the direction of degree for the neighbors. The above equations account for the weight of each link, given by a.

For undirected networks, we simply use the degree of each node for the purpose of calculation. Thus, the equation for degree correlation for undirected networks is as follows:

$$k_{nn,i} = \frac{\sum_j a_{j,i} k_j}{k_i} \tag{5.14}$$

Table 5.4 lists the basic network theory metrics discussed in this section and gives the formula used to calculate each.

5.2 MODELING NETWORK DYNAMICS

The measures of network theory discussed in the previous section are useful for analysis of static networks. That is, if we know the set of nodes and links of a network, we can use the above measures to gain insights on the network's topology and begin to identify properties and behaviors of the network. In terms of application of network theory to systems engineering, using a static model of a system as a network, we can identify and analyze the interactions among its components, and use this analysis to predict system-level behaviors as a function of their component behaviors and their interactions.

We now wish to discuss the topology and behavior of networks not just as they currently exist, but also how they will evolve with time. Comparing a network's

Table 5.4
Basic Network Theory Metrics

Metric	Formula	Comment
Degree	$L=(1/2)\sum_{i=1}^{N} k_i$	Undirected graphs
	$L=\sum_{i=1}^{N} k_i^{in}=\sum_{i=1}^{N} k_i^{out}$	Directed graphs
Average degree	$\langle k\rangle = 2L/N$	Undirected graphs
	$\langle k^{in}\rangle = \langle k^{out}\rangle = L/N$	Directed graphs
Clustering coefficient	$CC_i = \frac{2L_i}{k_i(k_i-1)}$	
Eigenvector centrality	$x_i = \lambda^{-1}\sum_j A_{ij}^{w} x_j$	λ is the largest eigenvalue
Betweenness centrality	$B_i = \frac{2}{(n-1)(n-2)}\sum_{s\neq t\neq i}\frac{\sigma_{st}(i)}{\sigma_{st}}$	
Assortativity	$k_{nn}^{in-dir}(k_i^{in}) = \frac{\sum_j a_{j,i}k_j^{dir}}{k_i^{in}}$	In-degree correlation
	$k_{nn}^{out-dir}(k_i^{out}) = \frac{\sum_j a_{i,j}k_j^{dir}}{k_i^{out}}$	Out-degree correlation

behaviors against the ones that we desire, and with a mapping between network topology and its behaviors, we can potentially influence the growth of networks to guide them toward certain topologies. In the context of SoS, this means that we wish to identify the natural dynamics of a complex networked system and guide future evolution of the SoS toward desired behaviors. Network theory can help us understand the dynamics of evolution of networked systems.

The most generic network growth algorithm can be represented as shown in Figure 5.5. This figure shows that starting from an initial network, we iteratively add or remove links over successive time steps based on a given logic until we arrive at the new network topology at a future time step. The difference among various algorithms lies in the middle box, i.e., the logic used in the addition or the removal of links. We will now discuss algorithms for growth of random networks and scale-free networks.

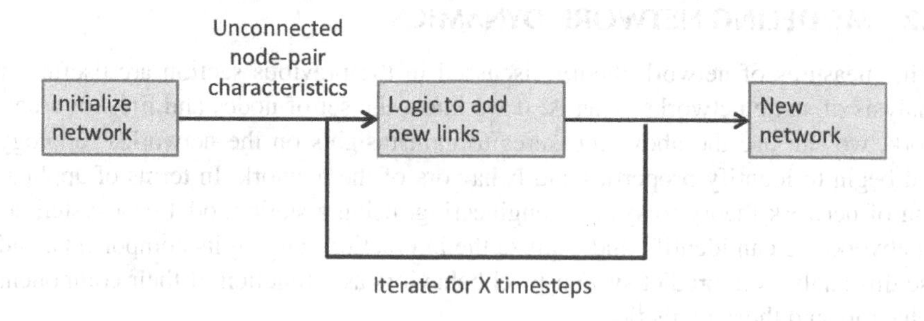

Figure 5.5: Generic network growth algorithm.

5.2.1 GROWTH ALGORITHMS: RANDOM NETWORK

Perhaps the simplest network growth model is that of random networks. A *random network*, as we discussed earlier, is one in which any pair of nodes has a probability p of being connected. There are two ways in which we can model the growth of random networks.

In the first version, called the Erdős and Rényi model, we start with a set of N nodes and L links. These links are then randomly placed within the network. In the second version, we do not know the number of links beforehand. Rather, we know the fixed probability p with which we connect two randomly selected nodes.

An interesting property of random networks is phase transition, where, below a certain probability of linking nodes $p < \lambda$, we have many small isolated networks while for $p > \lambda$, all of the small networks link together to make a single giant component. Here, λ is the critical point.

The degree distribution in a random network is given by the binomial distribution, though it can be approximated by a Poisson distribution for large number of nodes. Though random networks are easy to study theoretically, most real-world networks are not random which makes this model of network dynamics of limited utility. In social networks, we observe that while some individuals are highly connected and may interact with lots of people, many other individuals have a rather limited network of friends and acquaintances. Similarly, a popular topology in aviation networks is called a hub-and-spoke network. We need other models of network topology which can better represent such features.

5.2.2 GROWTH ALGORITHMS: SCALE-FREE NETWORK

A network model which better represents degree distribution in real-world systems is one in which a few nodes have very high degree while a large number of nodes have very low degrees. The degree distribution of a scale-free network is given by:

$$p_k \sim k^{-\gamma} \tag{5.15}$$

where γ is the degree exponent.

The Barabási-Albert algorithm of network growth follows the preferential growth pattern and results in development of networks with scale-free degree distribution. Briefly stated, in this algorithm, new incoming nodes prefer to attach to existing nodes in proportion to their degree. In other words, a new node will prefer to attach to a node with high degree, a hub, over one with a low degree. Suppose an existing network has a total of N nodes and a new node is added. The probability of this new node linking with an existing node i is given by

$$p_i \sim \frac{k_i}{\sum_j k_j} \tag{5.16}$$

where k_i is the degree of node i. Thus, a new node is more likely to link with an existing node with high degree as compared to an existing node with low degree. Over time, this leads nodes with high degrees to gain even more neighbors resulting in a "rich-gets-richer" phenomenon.

5.3 USING NETWORKS FOR SOS MODELING AND ANALYSIS

The complexity of a system results from the number of its subsystems, i.e., its size, the complexity of individual subsystems, and the complexities of subsystems' interactions. We can work on SoS analysis through the following three questions:

1. What are its constituent systems?
2. What is the structure of their interactions?
3. What are the resultant behaviors of the assembly?

The first one of these is what we addressed in the definition phase. With networks we answer the following two. For the purpose of modeling an SoS as a network, we identify the right set of nodes and links at the appropriate level of abstraction to answer our questions, and we map the observed structure to the observed or intended behaviors.

Given the task of engineering an SoS, we face the challenge of selecting the right mix of systems to assemble alongside the set of interconnections that link together each of the individual systems. The topology of interactions among components of a system has a huge bearing on the system's behaviors [13]. Features of system structure such as a hierarchy of components and the presence of feedback loops due to both direct and indirect directional links among systems affect how the whole behaves differently from a simple aggregation of its subsystems, leading to the observed emergent behaviors. Using networks to represent SoS, we can abstract away the details of individual systems and focus instead on the graph theoretic measures and dynamics of complex systems.

Network theory can be used to answer a variety of questions including, "Which vertex in this network would prove most crucial to the network's connectivity if it were removed?" for a small sized network by the use of metrics like node centrality and connectivity, or "What percentage of vertices need to be removed to substantially affect network connectivity in some given way?" for larger networks [136]. Especially for large-scale complex networks, such as the ones we are interested in, both the availability of tremendous amounts of data and the continuously increasing computational power to analyze this data further add to the utility of network science as a tool for systems' analysis. The tools of network theory enable us to make inferences about a system even when we cannot actually look at it.

Thus, we want to represent and analyze the architecture of an SoS, specifically, the network of interactions among the constituent systems because most of an SoS' emergent behaviors can be traced to the interdependencies among its constituent systems. When developing network models of an SoS, we need to address a number of considerations:

1. Type of Situation: Where is the network topology coming from? Empirical vs. Exploratory Modeling; Analytical vs. Simulation
2. Mappings between topological characteristics and operational performance metrics: average clustering coefficient does not mean anything to a FAA decision-maker, a military commander, etc.; Use correlation charts, but also offer evidence of causation

3. Evolutionary Scenario Generation: Dynamics of Networks – How and why might topology change?; Dynamics on Networks – How and why does a process propagate on the network?

Figure 5.6 visually represents how networks fit in to the SoS analysis process. We start by obtaining data from either observations made in the real world or from simulating available models of the system. From this data, we identify components and their interactions and represent them respectively as nodes and links of a network. Having done this, system analysis will follow guided by our objectives and problem context and scope. For example, we can assess system complexity by the analysis of the structure and prevalent patterns of interactions using any of the network measures discussed in Section 5.1.2. We could compare different structures of interactions and correlate them with system performance to predict or guide future topological evolution by using network growth algorithms such as those discussed in Section 5.2.

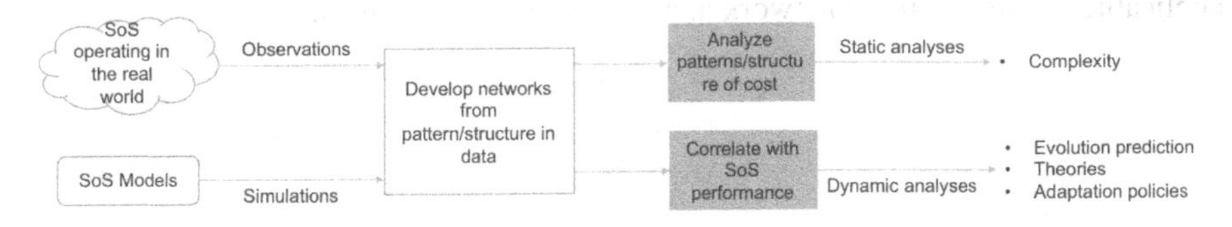

Figure 5.6: Using network theory in SoS process.

Due to their versatility, networks have been used for modeling and analysis of various complex systems including social networks, telecommunication systems, transportation systems, etc. For example, network theory has been used to map complex system architectures to their performance [13], and to study connectivity within the air transportation system [48]. In the latter example, connectivity metrics for the air transportation system are useful to explain features such as the capacity of the links and nodes, the network robustness, and even their evolutionary characteristics. Networks have also been used to represent interactions among humans including when these interactions have a hierarchy [144, 169].

5.3.1 MODELING INTERACTIONS AND FLOW WITHIN NETWORKS

We identified four types of SoS (see Section 5.1) based on the level of control available to a central authority. However, an SoS may not just be a single network, rather it can be a collection of multiple subnetworks of systems at different levels of hierarchy. Based on the division of decision-making authority, we can classify networks as centralized, decentralized, or distributed.

Centralized This is a type of system with a single top-level authority which exerts control, perhaps via multiple layers of hierarchy, over all subsystems that form part of the whole.

Decentralized This is a hybrid of centralized and distributed network of systems with multiple large nodes (called hubs) and layers of hierarchy. A federated System of Systems is a type of decentralized network of systems with higher level of autonomy of individual systems.

Distributed This is a type of network of systems which can be represented as a grid or a mesh. In a distributed system, several subsystems link together to enable its capabilities. For example, peer-to-peer networks are a type of distributed network of systems.

Figure 5.7 compares these three types of systems and Figure 5.8 distinguishes between the notions of distribution and decentralization in SoS. We can see that a decentralized system is hierarchical where a node controls other nodes in layers directly below it and the nodes within the same layer may not coordinate. In contrast, a distributed system is non-hierarchical where nodes are equal and usually cooperate with their neighbors using common protocols. The notion of distribution is, therefore, applicable to every disjoint network at a single level of hierarchy.

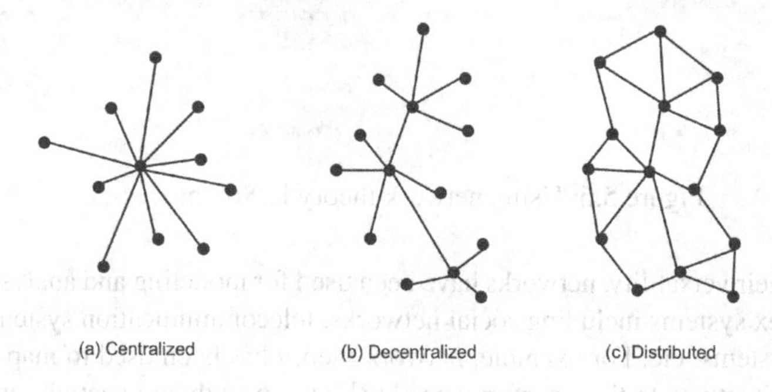

Figure 5.7: Comparison of centralized, decentralized, and distributed forms of network architectures.

In network-theoretic terms, a centralized system is a type of scale-free network with one hub and many spoke nodes. Similarly, a decentralized network also has a hub-and-spoke topology with multiple hubs and a number of spoke nodes. A distributed network, on the other hand, is a random network and most of its nodes have similar degrees.

Since humans are invariably an intrinsic part of complex SoS, analysis of social networks within such systems address issues of centrality (which individuals are best connected to others or have most influence) and connectivity (whether and how individuals are connected to one another through the network). We can identify subnetworks of component systems which are connected to one another with a more dense set of links than they are to other systems. Within social sciences, such subnetworks are called communities.

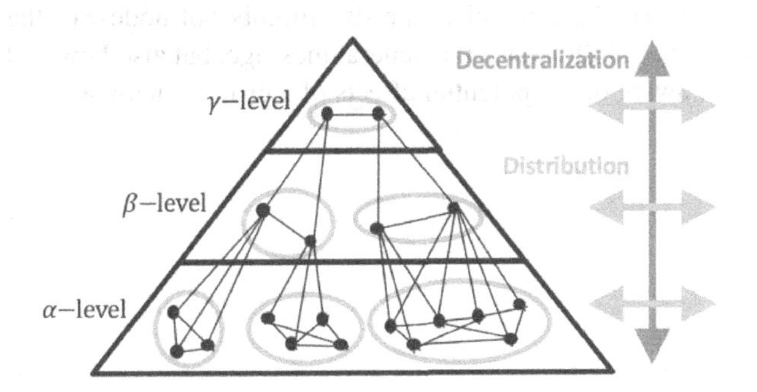

Figure 5.8: Decentralization and distribution define different aspects of a network. Decentralization is across levels of hierarchy, whereas distribution is applicable to multiple disjoint networks at the same level of hierarchy.

5.3.2 BEHAVIORS OF COMPLEX NETWORKS

Small-world effect: If the number of vertices within a distance r of a typical central vertex grows exponentially with r – and this is true of many networks, including the random graph – then the value of l (the mean geodesic distance between vertex pairs) will increase as $\log n$ (n: number of vertices). Networks with power-law degree distributions are sometimes referred to as scale-free networks, although it is only their degree distributions that are scale-free.

5.3.2.1 Example 4: Disease Spread via Random Geometric Graphs

Let us look at an example to see how networks can help us derive system insights. Random Geometric Graphs (RGG) are graphs in which each vertex is assigned random coordinates in a geometric space of arbitrary dimensionality and only edges between adjacent points are present. Here's how RGGs can be used in study of disease spread:

- Placing vertices at random uniformly and independently on the region
- Connecting two vertices, u, v, if and only if the distance between them is at most a threshold r, i.e., $d(u,v) \leq r$
- Can obtain properties of graph as a function of r

RGGs can exhibit features of both regular lattices and random graphs. At low values of domain-circle radius, the contact-network resembled a lattice, and diseases do not spread effectively on lattices since infected individuals mostly interact only with other infectives (Watts, 2003). As the domain-circle radius was increased, the network topology resembled a random graph with higher probability of reaching the epidemic threshold.

More generally, there are some general insights the community has developed that appear to apply in practice (in large measure due to common sense). One example is the simplest of network characteristics: the size (number of nodes) of the network. The graphic in Figure 5.9 illustrates the general message, but also how and why SoS architects must be aware of the potential effects of a growing network.

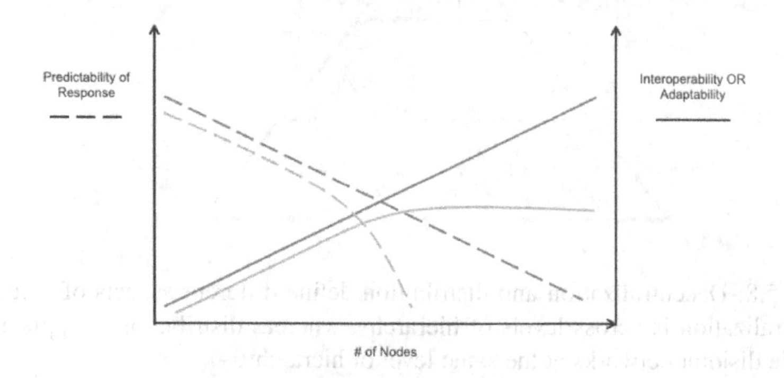

Figure 5.9: In general, as network size (here simply represented by number of nodes) grows, the predictability of network behavior diminishes. However, the benefit (and motivation) to increase network size is the adaptability derived from more inter-operating nodes. The nominal forms of these relations are shown in blue; however, an undesirable form (an SoS design to avoid) is indicated in orange, where predictability falls precipitously, while adaptability benefits plateau.

Eventually, we need to be able correlate (map) topological characteristics to performance metrics that is applicable to the problem. The network measures we have discussed in the previous section do not mean anything directly to the decision-makers. Instead, we should be able to describe the actual meaning of network measures in problem. This translation to relevance was part of our journey, in particular in work with the Federal Aviation Administration (FAA) that, after a number of iterations, in fact uncovered important ways to correlate network structure to relevant performance.

5.4 MODELING THE AIR TRANSPORTATION SYSTEM USING NETWORK THEORY

The air transportation system is a complex SoS comprised of resources such as airports, aircraft, and other infrastructure, along with many different stakeholders each of whom operates independently to fulfill its own objectives. The definition of this system using the DAI framework was presented in Chapter 3. In this section, we present an example of application of network theory to a real-world SoS, the air transportation system.

To model this system as a network, we represent the airports as nodes connected by links which represent direct flights between them. Alternatively, if we use links to represent flow of passengers, then, in general, a link would connect more than two nodes whenever a passenger passes through more than two airports such as in case of a layover. The links in this case would be hyperedges which represent passenger itineraries, and the graph will be classified as a hypergraph.

The aviation system can be modeled as a network of links which indicate direct flights. In fact, different airlines may have different network topologies corresponding to their business models. For example, low-cost airlines frequently have more direct flights, which makes their network closer to a random network. The legacy airline model, however, is closer to a scale-free network, which leads to the hub-and-spoke network of many of the large airlines.

We may instead want to study just one airport. In such a case, the nodes would be the different subsystems within an airport and the links would be their interactions. Thus, nodes can be represented with agents with their own behaviors. Similarly, links would have their respective attributes depending on the system we wish to model.

Depending on our problem, therefore, the air transportation system can be represented as many different types of networks. We can also choose to represent the air transportation network as a hierarchical system of networks of multiple stakeholders, as done in [132]. In this study, the authors modeled the future evolution of the aviation network as a result of decisions of two different stakeholders, viz., airlines and passengers. Figure 5.10 shows a hierarchy of the networks of these two stakeholders.

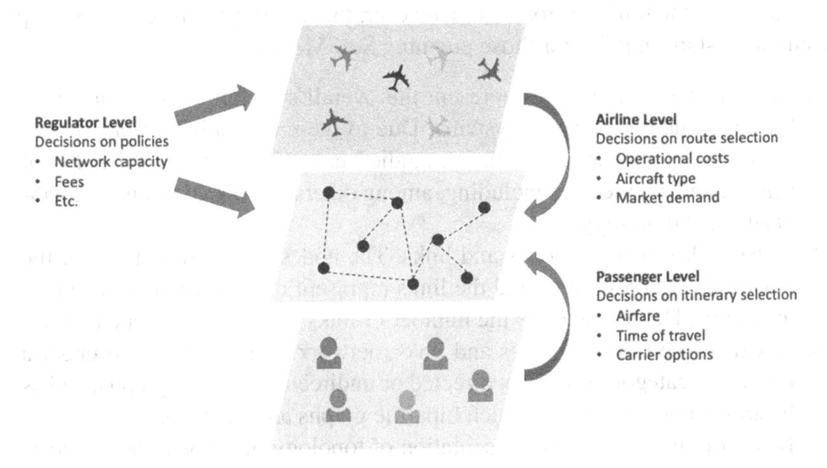

Figure 5.10: Different stakeholders and resources can be organized in their own layers as networks to represent the air transportation system [132].

DeLaurentis et al. [48] have given an example study on use of networks for air transportation system analysis. They model the US Air Transportation System (ATS) using data from 2004 on the principles of Network Theory. In order to study the US ATS, they use such global measures as network density, degree distribution and average shortest path length and local measures such as strength of a node, betweenness centrality and clustering coefficient.

The ATS at the national network is of the scale-free kind of topology, i.e., there are many different airports connected to few others, while there are a few airports that are connected to many others. This is the hub-and-spoke concept of air transportation.

The same can be said for the airline networks with the networks of Delta and United Airlines, for example, following the scale-free topology very closely; the Southwest network, on the other hand, has some properties of random topology. In addition to the above, studies were conducted at the airport level to determine the importance of individual airports. In this case, four different measures were used to compare the airports resulting in a different outcome in each case — including that of Anchorage airport being very important on the betweenness measure.

The presented method of analyzing the ATS requires minimal set-up effort and can give useful insights, and hence it can be useful in studying similar such networks. However, it can still be developed further to include factors like technology and policy changes and the ability to predict them. Thus, rather than using this method only to measure the network characteristics, it can be developed to predict future scenarios which will help in designing networks that are more robust and efficient.

5.5 CHAPTER SUMMARY

In this chapter, we introduced network theory as a powerful approach to modeling and analyzing complex systems, especially SoS. The study of networks is a mature yet still evolving discipline, and the references cited in this chapter provide a more in-depth theory on the topics introduced. However, the theory presented in this chapter is a sufficient starting point for those pursuing SoS M&A.

1. Graphs provide a way to represent the overall structure and relationships between many real-world systems. Due to their versatility, graphs, called networks in network science, are studied extensively in several fields of science and engineering including, among others, the social sciences, mathematics, and biology.
2. A network is a set of nodes and links. The nodes can represent any of the independent SoS entities and the links represent their relationships and interactions. Degree refers to the number of links a node is associated with.
3. Based on the types of nodes and links, networks can be classified under a myriad of categories, e.g., as directed or undirected, as hypergraphs, and as heterogeneous graphs, of which bipartite graphs are one type.
4. Based on the statistical representation of topology of links in the network, we classify *random* networks as those in which any two nodes connect to each other with a probability p which produces a Poisson degree distribution, while *scale-free* networks are those which follow the power-law degree distribution and exhibit some kind of "hub and spoke" topology.
5. Once defined, we can analyze a network using any of the large number of metrics to understand a network's characteristics such as its topology, complexity, etc. Examples of network metrics include node degrees and degree distribution, clustering coefficients, and centrality measures including betweenness centrality and eigenvector centrality.
6. Network theory can help us understand and predict the properties and behaviors at different hierarchical levels of the SoS on the basis of measured structural properties and the local rules and incentives governing individual components.

5.6 DISCUSSION QUESTIONS AND EXERCISES

5.6.1 DISCUSSION QUESTIONS

1. Find an example of an undirected network in the real world. Provide justification as to why this network operates as an undirected network.
2. Define a "cycle" in a network. Describe a cycle assuming a fully connected network with nodes numbered 1, 2, 3, 4, 5, and 6.
3. In your own words, explain why assortativity could be important to System of Systems theory.

5.6.2 EXERCISES

1. Design an undirected network with N = 7 and L = 12. Based on how you drew your network, classify it as either fully connected, random, or scale-free. Justify your decision with a short paragraph response.
2. Using the same graph from the question above, create an adjacency matrix that defines relationships within the graph.
3. Design a directed network that has N = 6 and L = 10. Choose directionality on your directed network. Once directionality is established, build the adjacency matrix associated to your graph.
4. Design a network with an average degree of 2.
5. For the following network (Figure 5.11), do hand calculations (show your handwritten work) to compute (a) for each node: degree, clustering coefficient, betweenness centrality, and degree correlation (all four permutations: In-in, in-out, out-in, and out-out), (b) for the network: degree distribution(s), diameter, average shortest path, and network average clustering coefficient (NCC). Note that this network is directed, so make appropriate consideration when doing your calculations.

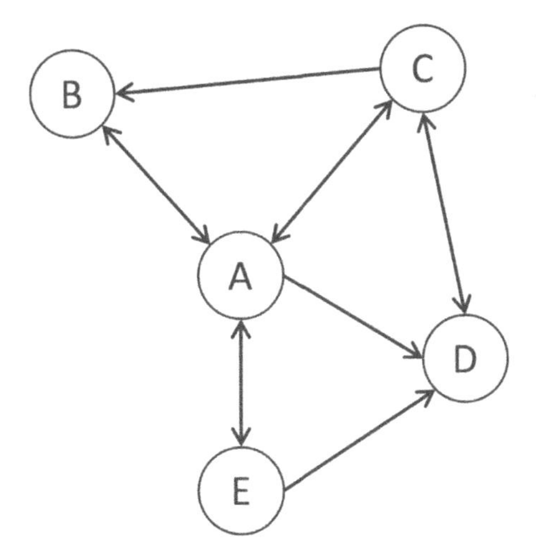

Figure 5.11: Figure for question 5.

6 Agent-Based Modeling

When we discussed the role of modeling and analysis in engineering, we highlighted the difficulty of decision-making for complex systems (see Section 2.2). Modeling can be a challenging task for large-scale systems, even when they can be approached from first principles such as by using laws of physics for physical systems or by fitting a model to empirical data. When we have good understanding of a system's dynamics, we start with *a priori* knowledge of its behaviors and use the process of deduction by reasoning logically about the implications of our existing knowledge to arrive at the desired model. When this approach is not feasible, we can attempt to use induction to discover patterns within available data. However, this requires that we either get access to real-world operational data or have the ability to generate data by experiments, and the richer the details we wish to model, the more data we need. Further, even if we were to somehow develop surrogate models of such systems using, for example, statistical approaches such as regression, we will obtain a picture of the aggregate behaviors without the ability to trace such behaviors back to those of the constituent elements.

Once we have the requisite mathematical models, we may face two different situations where we can solve these models analytically or we may not have any analytic solutions, in which case we can use numerical approaches. A third, even more complicated situation is when we do not have any mathematical models available, as is usually the case with large-scale complex systems. In general, in cases where we either do not have reliable analytic solutions of the system under study or we lack mathematical models of the dynamics of the complete system, we try to derive insights using simulations. Agent-based modeling (ABM) is a simulation technique which provides a third way, that of a "plug-and-play" approach where we model the constituents and see what happens naturally as a result of their interactions.

In this chapter, we will give a brief description of what ABM is and how it is useful in SoS modeling and analysis problems. We will discuss how to model individual constituent systems of our SoS as independent decision-making entities and observe behaviors of the whole by simulating their interactions. We begin by discussing what agents are and how they can be classified and modeled. Following this, we discuss when ABM is useful and how it can be used in the context of SoS problems. We will conclude this chapter with a discussion of an example of ABM application to the ATS problem.

6.1 A BRIEF INTRODUCTION TO AGENT-BASED MODELING

Agent-based modeling (ABM) is an approach to simulating the actions and interactions of a collection of agents, i.e., systems, to observe their effects on the whole system. Rather than identify the dynamics and models of the whole system, we can program each individual component and then simulate the behaviors when they interact with one another. Then, by observing patterns of interaction and their effects on system-level dynamics, we can derive insights of large-scale behaviors that would be otherwise difficult to obtain.

An ABM is especially useful for modeling complex adaptive systems, because it is a dynamic model that represents both the individual agents and their collective behavior. Usually, even simple rules of behavior applied uniformly to all agents in the simulation can produce surprisingly complex dynamics. To see the complexity of outcomes that can be observed even with simple rules, first consider Finite State Automata (FSA). An FSA is a computational machine with a finite number of possible states, and this machine can be in exactly one state at any given time. The machine can transition through the set of discrete possible states based on inputs which trigger certain conditions. The specification of an FSA consists of five parts:

1. A set of inputs: $A = \{a_1, a_2, \ldots, a_M\}$
2. A set of states that the system can be in at any given time: $S = \{s_0, s_1, \ldots, s_N\}$
3. A set of goal states: Y
4. A set of initial or default states: s_0
5. A state transition function which defines the rule for change of state: $F : S \times A \rightarrow A$

As an example, consider the sequence of states of an aircraft during departure; Figure 6.1 shows a representation of this sequence in form of an FSA. The starting state is when the aircraft is at the gate. The pilot requests for permission to push back. If no clearance is granted, the aircraft remains at the gate, whereas on receiving approval, the pilot initiates push-back to the ramp. Then the same sequence of operations repeats with the pilot first requesting permission to taxi followed, upon arriving at the runway, by departure clearance. If the departure proceeds nominally, the aircraft departs the airport; otherwise, if there is a deviation from the flight path, the aircraft moves in to a separated state wherein the pilot initiates recovery until return to nominal. Eventually, the aircraft departs the airport, which is the goal state. This is a basic representation in which the aircraft passes through the states of waiting at the gate, waiting on the ramp, waiting at the runway, and departure process, along with a deviated state which corresponds to an off-nominal condition [9]. In each of these states, a pilot conducts multiple activities, and an even more detailed FSA can be modeled by accounting for all pre-departure activities of a pilot.

The above specification is for a deterministic FSA; for a non-deterministic FSA, the transition function will change to denote probabilistic state transitions. For example, imagine a single cell which can be in one of two states, viz., "alive" or "dead" with a certain probability p. At each time step, we generate a random number between 0 and 1, and the cell either comes alive or stays alive if the random number exceeds p; otherwise, it dies off. In this model, the probability p defines the state transition function, the cell is randomly initialized in one of its two possible states, and there is no goal state.

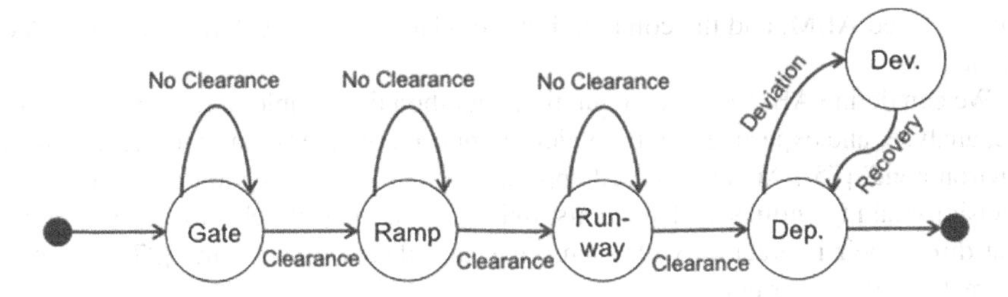

Figure 6.1: Representation of aircraft departure sequence as an FSA. An aircraft moves through the states of Gate, Ramp, Runway, Departure, with an off-nominal condition marked as Deviation.

Though an FSA models several simple real-world machines such as an elevator, a vending machine, among others, let us extend the above example of a cell transitioning between dead and alive states and create a more elaborate implementation of the same idea of cellular automata; this is a set of cells in a grid, where each cell is an FSA. A cellular automaton adds complexity to the above model by simulating interactions among multiple cells and generating complex outcomes as a result of those interactions.

The most famous example of a cellular automaton is the Game of Life [68]. The Game of Life consists of a 2D grid of cells each of which can be in one of two states, alive or dead, and has eight neighbors: north, south, east, west, and four diagonals, called a "Moore neighborhood" [54]. At the beginning of the simulation, each of the cells is randomly initialized as either alive or dead. Thereafter, in each time step, the next state of each cell depends on its current state and its number of live neighbors. If a cell is alive, it stays alive if it has two or three neighbors who are also alive, and dies otherwise. If a cell is dead, it stays dead unless it has exactly three neighbors who are alive.

The above specification of Game of Life is one of many possibilities that exist; other forms differ in the rules of state transition. However, even with simple rules of transition and limitations on available information (cells can only observe the behaviors of immediate neighbors and have no memory), rich behaviors can be modeled. If we start a Game of Life simulation from a random starting state, a number of stable patterns emerge, some of which have been named [54]. Therefore, two key features of this simulation are that the rules of state transition are simple and all decisions are locally made based on local (neighborhood) information.

While the Game of Life applies identical rules to all cells in the grid, our interest lies in the ability to model a complex system at various levels of detail within its subsystems, i.e., we want to model assemblies of heterogeneous systems. We have emphasized that a major hurdle to such an analysis is that we lack any reasonable model of such large aggregates of systems [14]. This brings us to the idea of simulating models of individual systems to fulfill our objective of analyzing the aggregate behaviors of large assemblies of such systems. The modeling approach based on this idea is called ABM, and the computational models of individual systems are called *agents*.

We can define ABM as "a computational method that enables a researcher to create, analyze, and experiment with models composed of agents that interact within an environment" [75]. An agent-based approach "employs a collection of autonomous decision-making entities, called agents, imbued with rules of behavior, often simple, that direct their interaction with each other, and their environment" [27]. ABM is characterized by four constructs:

Agents: The simulated behavioral entities of various types. These are the main decision-making entities in our model.

Objects: The set of all passive entities that agents interact with and which form a necessary part of our model.

Environment: The topological space where agents and objects are located. The environment is also the source of resources needed for agent operation.

Communications: The set of all possible interactions between entities, i.e., the set of information, material, and energy exchanges among the entities in the model.

In the following subsection, we will describe agents, their types, and how they are modeled.

6.1.1 WHAT ARE AGENTS?

Agents are discrete entities that function independently of other agents in their environment. The boundaries of an agent are clearly defined, which means that we can classify any component as either belonging to or being outside of the agent. For our purposes, we will assume that any system is an agent if it obtains inputs by sensing its environment, makes decisions in accordance with its goals and preferences, and generates outputs through its actions. Systems which lack these features, especially decision-making capability, are classified as objects, which, though essential to the models, are not the focus of our efforts.

Figure 6.2 shows a schematic of an agent. Every agent acts within an environment from which it obtains inputs using its sensors. Using these inputs, the agent updates its current beliefs and knowledge of the state of the world, which it then combines with its preferences (its desires and objectives) and, depending on the current state, makes decisions about future actions. Specification of the agents' sensing, decision-making, and acting capabilities is a key part of an agent model.

We refer to the form of agent model shown in Figure 6.2 as the Behavior-Knowledge-Intention (BKI) (or BDI, which stands for Behavior-Desire-Intention) model. In this model, agents are computational entities that have three attributes: behaviors, knowledge (or desires), and intentions. Behaviors refer to the agents' capabilities, i.e., the actions they can take in accordance with their preferences. For example, movement from one point to another predetermined point when set in motion is one possible behavior of a robot. On the other hand, for a computational agent, generating outputs in response to inputs will be its behavior. Specification of agent behaviors is a source of classification, which we will discuss in the next subsection.

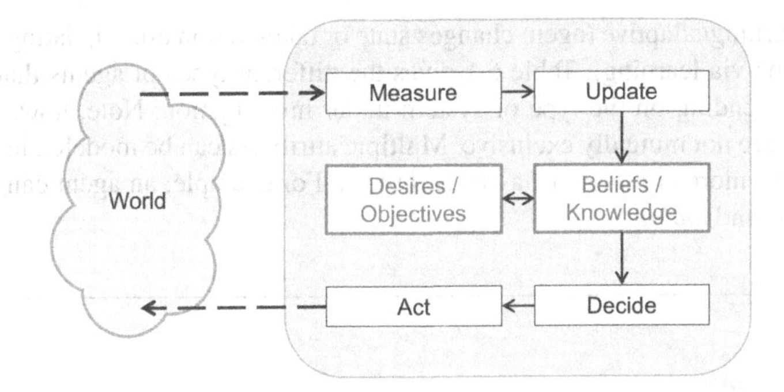

Figure 6.2: The Behavior-Knowledge-Intention model of agents.

Knowledge (or belief) is everything that the agent knows including its own desires or goals as well as its own current state and that of the environment. Specification of agents' knowledge is a key activity in modeling. The knowledge that is encoded by the modeler prior to operation is referred to as *offline* information. For agents with an ability to learn, knowledge can change over time during simulation. In such cases, an agent obtains information during run time by sensing its environment or by exchanging information with all other agents that it interacts with, and updates its knowledge base to reflect the newly acquired information. This knowledge is called *online* information. Regardless of whether knowledge was acquired online or offline, the agent uses the sum of its knowledge to guide decision-making. Sometimes, the agent may not be able to sense its environment exactly as it is. Rather, its input may include noise. In such cases, the agent will need to make the most of available information when making decisions. The agent does so by maintaining models of its world which is revised as new information is acquired. We can refer to an agent's internal models of the world as its belief.

Finally, intentions are the agent's short-term plans for action. We need to distinguish between an agent's long-term goals and desires, which form part of the agent's knowledge base, and its intentions which result from decision-making in response to inputs from the world. Intentions encompass the agent's preferences and its end goals with regard to its decisions or the potential outcomes of the world, i.e., the state of the world. Autonomous decision-making agents especially make use of their intention to obtain maximum benefit for themselves while they interact with the world. This notion is described in economics as "rational" behavior of an agent. In fact, we assume that an agent is a willing participant in system operation so long as it finds it beneficial to do so. This is particularly relevant to SoS, where, in general, component systems may have the freedom to either be a part of the whole, or not.

6.1.2 TYPES OF AGENTS

We have a lot of flexibility in defining agent functionality, with behavior types ranging from simply reactive (agent changes state or takes action based on fixed rules) to learning/adaptive (agent changes state or takes action after updating internal logic schema via learning). Table 6.1 gives the different types of agents that can be modeled depending on the type of system under investigation. Note, however, that these types are not mutually exclusive. Multiple attributes can be modeled in a single agent in case more complex behavior is desired. For example, an agent can be both *autonomous* and *mobile*.

Table 6.1
Types of Agents

Type	Description
Autonomous	Runs without continuous user input
Interface	Requires assistant user
Info-gathering	Collects, filters, and classifies information
Goal-based	Does not have a set solution path, and does not care about a utility value
Utility	Does not have a set solution path but cares about a utility value
Reactive	Has a set solution path but cannot change its behavior based on past experiences
Adaptive	Can change its behavior based on past experiences
Mobile	Can move

Figure 6.3 shows a flowchart of questions that can be asked for selecting the type of agent desired. Let us emphasize again that an agent need not, and, in general, will not belong to just one type. Thus, for example, an "info-gathering agent" is an agent that senses its environment or receives information from other agents in its environment and acts on it. Similarly, an "autonomous agent" is an entity that functions continuously and autonomously in an environment without user input; however, an autonomous agent will generally need to be info-gathering as well since it will need to sense its environment during the course of operation.

6.2 THE WHEN AND WHY OF ABM

Agent-based modeling (ABM) is frequently used to model complex adaptive systems, which are systems whose components change in response to stimuli from the system's environment, and the properties of the whole system emerge from the actions of its components. Using the traditional approach of seeking analytical solutions of rigorous mathematical models of such systems is an especially challenging undertaking, and may not even be feasible in cases where we lack an understanding of the system's dynamics. Earlier in this chapter, we identified three distinct possibilities: when we have mathematical models available along with analytical solutions for them, when we have mathematical models but no analytic solution methods, in which case, they may be solved numerically, or when we do not have any

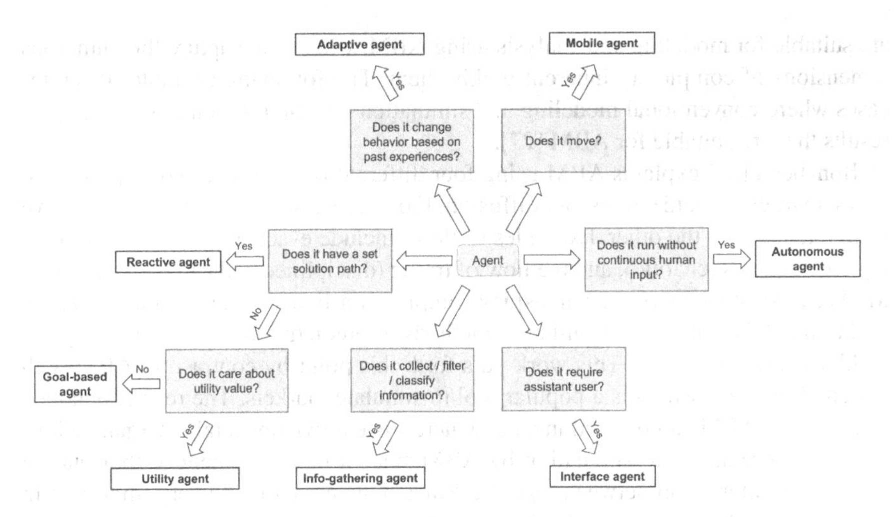

Figure 6.3: Selecting the type of agent.

mathematical models. The ABM approach can be used in all three cases, though it is particularly useful in the latter two [12]:

1. When analytical solutions to problems exist and can be obtained, agent-based computing can serve as a check on the results,
2. When analytical solutions exist but are difficult to solve, agents can give direction of progress and some idea of the solution structure, and
3. When the solutions cannot be obtained analytically, agent-based computing may be the only way forward.

Thus, the choice of when and how to use ABM depends on the problem structure. In cases where we have both system models and analytical methods of solution, ABM can either help validate the models or provide direction for further exploration. Analytical solutions to problems are generally based on a set of simplifying assumptions, and while such assumptions make it easier to obtain results, they usually decrease the fidelity of such solutions [100]. ABM, on the other hand, can be used to model the interacting 'elements' of a problem and then run to solve the problem with little or no assumptions involved. This puts the agent-based method at an advantage over conventional approaches.

ABM is even more useful, and perhaps the only option, in the third case where we lack either system models or analytical solutions. The presence of well-established laws of physics, high interdependence of system components, and/or uncertainty in governing equations are reasons not to use ABM, while, when factors such as learning and adaptation (generally found in humans and animals) or operation under uncertainty of economic, social and political factors are present, then ABM may be used. For example, social science problems have inherent non-linearities which make them difficult to model using traditional analytical techniques. Thus, these problems are suitable for modeling and analysis using ABM, which can capture the numerous dimensions of complexity inherent within them. The following example illustrates cases where conventional modeling and simulation techniques would not give good results that are suitable for ABM [27].

Bonabeau [27] explains ABM using four different types of real-world problems: flows, markets, organizations and diffusion. Flows are cases where the entities move from one place to the other. Examples of flows include evacuation of buildings during emergencies (chaotic), and the flow of traffic (disciplined and can be controlled). Markets, like stock markets, can display complex non-linear behavior, and as a result traditional differential equation-based methods of predicting their behavior may not yield satisfactory results. This work puts forth this point by contrasting ABM with system dynamics, which is a popular tool to simulate markets. The results obtained explain how ABM can be used in cases where system dynamics fails. Organizations are even more suited to simulation by ABM since a major element of their nature arises from interaction between 'agents.' Since human factors are very important in determining behavior of organizations, most other techniques may not give comparable results. Finally, diffusion relates to spread of ideas or products in the society, a very important area of concern for business organizations. In short, ABM can be useful in predicting or studying those aspects of systems that have inherent non-linearities and hence cannot be modeled by traditional analytical techniques.

ABM is distinct from traditional modeling techniques because, using ABM, systems engineers can investigate alternative architectures and gain an understanding of the impact of the behaviors of individual systems on emergent behaviors from bottom-up rather than top-down as is done in other techniques. Once set up, an agent-based model can be simulated multiple times with various inputs and parameter values. Recurrent patterns of system behavior can then be recognized, especially those that are robust to variations in model configuration or inputs, and these can, in turn, be used to derive insights about the most fundamental properties of the system.

However, this inability to conclude anything based on just one run of the problem – inferences can only be drawn after repeated runs of the simulation – is also ABM's disadvantage. This is true despite the advances in computational facilities at our disposal which enable the use of ABM for simulation. Since ABM models constituents of the system at a lower level of hierarchy, the potential time required for simulation can be very high, making it uneconomical to use ABM for decision making. ABM's advantages of ease of formulation, scaling, and implementation of the problems may also be overwhelmed by the need to develop models for each problem at hand – a single generic model for use for all problems does not exist. As in all modeling, it is imperative to properly calibrate parameters of the agent model to ensure observations correlate to the physical interpretation of the model outcomes. This task can be addressed using available data from the real world.

While ABM can be a powerful method of analysis, it is not well suited to every problem. As we have noted, the presence of well-established laws of physics and associated governing equations is reason for leaving the ABM 'hammer' in the toolbox. On the other hand, when factors such as learning and adaptation (generally found in biological systems... like us!), operation under uncertainty, or economic, social and political factors are present, then ABM may be useful. For example, in a social science context, the network of agents affect each others' preferences or behaviors rather than actual decisions, which can be simulated to observe the dynamics of group formation and breakage [12].

Despite the disadvantages, some of which we have noted above, ABM is still an important tool for the study and architecture of SoS. Using this approach, we investigate alternative architectures and gain an understanding of the impact of the behaviors of individual systems on emergent behaviors. In most cases, ABM may be the only method of ensuring a reliable simulation. Further research and development of this technique is therefore imperative. One area of concentration at this stage is adding fidelity to the agents, which in some cases are intended to replicate humans, which are inherently irrational agents. However, absent sufficient knowledge about the dynamics and emergent effects of a situation, ABM may be the only means of modeling real-world problems with adequate fidelity.

6.3 AGENT-BASED MODELING FOR SYSTEM OF SYSTEMS

When describing the abstraction phase of the DAI process (Section 3.4), we identified three different classes of design variables that need to be defined for developing and utilizing a model: composition, configuration, and control. Composition variables answer questions of which systems to model. The ROPE table directly helps us select composition variables – in the context of ABM, we identify which resources and stakeholders to model as agents. Configuration variables define which operational inter-dependencies and constraints influence behaviors of the constituent agents. These variables identify the networks and rules of interactions. Selection of the agents' operational environment, including available resources and participating stakeholders, is part of defining model configuration. Finally, control variables are related to decision-making of the agents. Questions such as what the agents' preferences and their incentives of operation are, how much autonomy they have, etc. are part of this definition. Table 6.2 shows the outcomes of each of the three phases of the DAI process in the agent-based modeling approach.

Let us emphasize that the ultimate goal of doing ABM is not to prove, but to gain insight on the processes that may emerge and evolve in a complex system. Doing so, we may potentially be able to discover consequences which may not be obvious, i.e., the system's emergent behaviors. *Emergence*, which we identified as one of the key distinguishing features of an SoS, is by definition a difficult phenomenon to detect during the design stage, though it may potentially be obvious during system operation. Because of this, emergence receives much well-deserved attention in discussions of SoS. The primary utility of ABM is its ability to study or predict emergent phenomena from the bottom up rather than from the top down, as is common in other techniques. Thus, ABM is most useful in cases where independent entities are to be modeled under conditions of complex, non-linear, and heterogeneous interactions.

This leaves us with a lot of flexibility (and responsibility) regarding modeling SoS using an agent-based approach. For example, we have the choice of selecting

Table 6.2
Composition, Topology, and Control through DAI Phases

Phase	Composition	Configuration	Control
Definition	Identify the resources and stakeholders to model as agents	Identify the operational interdependencies and feedback loops among agents	The operational concept describes the division of control among agents
Abstraction	Define agents' desires and capabilities	Define the rules of interaction	Define the behaviors and control algorithms of each agent
Implementation	Develop agent models	Set up data flow (information exchange) among agent objects	Simulate agent model and analyze results

the level of fidelity of agent models. We will need to make decisions about what knowledge to encode within agents, what preferences to provide to each agent, and the behaviors the agents will have. For example, different sources place different levels of emphasis on higher-level cognitive behaviors such as ability to learn and adapt behavior when defining an agent [119].

A prudent approach in modeling is to abstract out the details of system components and start with a basic model with few defining parameters, before adding details. Just as we discussed increasingly more sophisticated models starting from Finite State Automata to the Game of Life, in order to simulate real-life problems more realistically, it may be beneficial to start with a core set of model parameters to formulate agent models, and then increase complexity of the model as needed. The easiest way to scale the problem could be to add more agents, though, to add more realism to the problem formulation, we may choose to make our agents more sophisticated by adding to their capabilities. Referring back to Figure 6.3, we can start with an agent model which fits neatly into one type, then add complexity by encoding additional behaviors. For example, we can set up an agent with a fixed utility function that controls its choices throughout the course of a simulation. As we gain confidence in the model and seek more capability, we can make the agent adaptive by allowing it to modify its utility function during run time based on information it gathers online.

One of the strengths of ABM is that it can accommodate heterogeneity in both agent behaviors as well as the network of their interactions. Thus, we need not model a uniform rule-set, and can rather observe a bottom-up evolution of the network that results from local decisions. ABM simulations can involve multiple agent types to replicate a certain environment with 'natural' properties. We can model interactions using networks, with rules or constraints which impose some structure on agent interactions based on data from the real world [23, 93].

Consider the concept of present-day, voice-based communications among participants of the air traffic management system. Figure 6.4 shows four participants along with communication links among them. To develop an agent-based model of these systems, we will first define agent objects for each of the four participants: airspace user operations (AUO), aerodrome operations (AO), air traffic control (ATC), and air traffic flow management (ATFM). Note that, in this case, these agents are not "tangible," meaning that they are not physical systems, rather they are services that enable operations in the present day ATS. Also, the links between these systems are of different types, particularly, voice links, data links, and human-machine links between these services and their human operators.

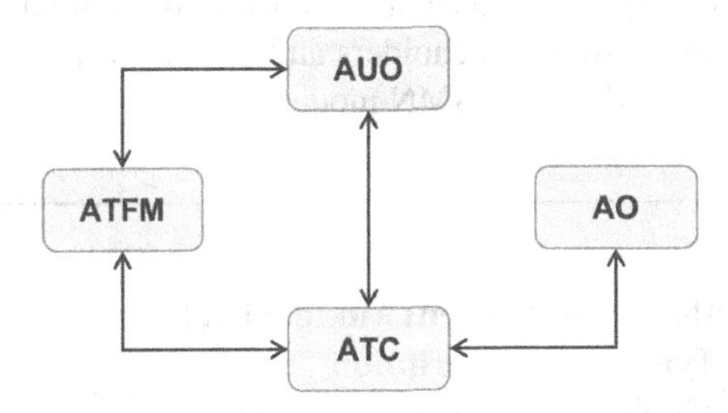

Figure 6.4: Voice-based interactions among participants in the air traffic management system [104].

We wish to move from primarily voice-based communication and control to trajectory-based operations in the future, which means an increasing emphasis on information sharing and coordination through services which enable sharing of data among participants. Our agent-based model of the system can help the transition from voice-based to trajectory-based operations by first developing agents of existing services, adding complexity to these agents and their interactions, and then developing and adding agents of future desired services.

6.4 EXAMPLES OF APPLICATION TO SYSTEM OF SYSTEMS

To demonstrate the use of ABM to SoS problems, we select a uniquely challenging design problem which has elements of development of novel technical capabilities, a distributed network of collaborating systems, and evolution over long time periods [151]. We are faced with a problem of defining a "Solar System Mobility Network (SSMN)" which is envisioned as a network of engineered systems and natural resources distributed throughout the solar system to support future long-distance missions. We will use ABM to model the constituent systems of this SoS and discuss how this model can help identify system-level dynamics and performance measures.

The DAI process model provides a structured approach to uncovering emergent behaviors by evaluating not just individual systems' performance, but also their interactions. ABM further supports such analysis by providing a flexible, bottom-up approach to uncovering the processes and patterns that produce emergent behaviors. Because we expect all constituent elements – drivers, disruptors, resources, and stakeholders – to contribute to overall emergence, we want to exploit ABM's flexibility to develop different classes of systems that interact with others while simultaneously exhibiting their own functions and capabilities.

Since the problem of SSMN design faces significant uncertainty with regard to desired capabilities and participating stakeholders and their objectives, the approach we take is to define multiple system architectures and simulate them to understand emergent outcomes. Here, architectures refer to different system configurations, meaning that an architecture is an instantiation of a set of agents and their connections. This means that we first identify the agents that form part of the system. Sindiy et al. [151] show agents of both the system stakeholders and resources. Table 6.3 shows the infrastructure agents included in the SSMN model.

Table 6.3
Infrastructure Agents in SSMN Agent Model [151]

Agent	Type	Description
Cargo transport	Mobile	Transports supplies
Service mission	Reactive	Provides services and maintenance to other assets
Supply product facility	Reactive	Produces supplies *in situ*
Asset manufacturing facility	Reactive	Constructs and launches other assets

Once we identify the agents, we define their interactions, which sets up both resource networks and stakeholder networks. These two types of networks, together with identification of drivers, which are functions that represent stakeholder values, and disruptors, which are events that affect the resource networks and drivers, together form one specification of system architecture. Thereafter, the following steps in agent-based analysis and design include simulation of the model and calculation of performance metrics (Figure 6.5).

6.5 CHAPTER SUMMARY

In this chapter, we introduced and discussed using agent-based modeling for SoS problems. ABM is a powerful technique which models constituents of a system as independent agents and simulates their behaviors and interactions, from which insights about the whole system emerge.

1. ABM is useful in all three cases: when we have mathematical models along with analytical solutions, when have mathematical models but no analytical solutions, and when we have no mathematical models.

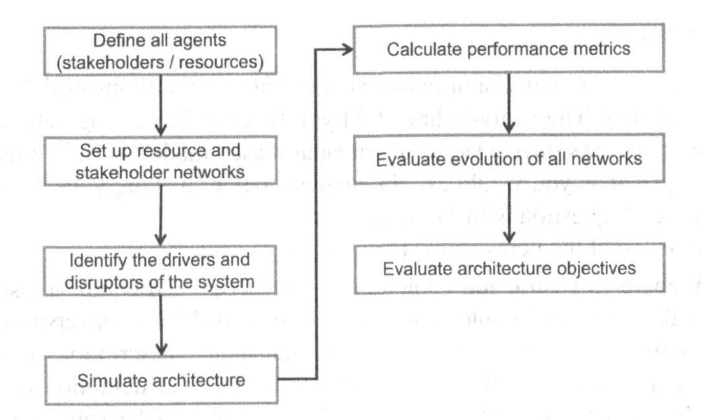

Figure 6.5: Agent-based approach for evaluating different SoS architectures.

2. An FSA is a computational machine which can only be in one of a finite number of states. The specification of an FSA includes a set of inputs, a set of states that the FSA can be in, a set of goal states, a set of initial or default sets, and a state transition function.
3. An agent-based model has four constructs: agents, objects, environment, and agent communications.
4. Agents are discrete entities which can represent any of the constituent elements of a system. They can be of many different types including mobile, autonomous, reactive, adaptive, etc.
5. A Behavior-Knowledge-Intention (BKI) model of agents is a useful framework for designing agents. Specifications of an agent's behaviors, its knowledge and intentions are a task of the modeler. Some agents may change their knowledge and intent during runtime.
6. For modeling SoS problems using ABM, we need to specify the composition of the SoS, configuration of interactions among agents, and control over agents decision-making and actions.

6.6 DISCUSSION QUESTIONS AND EXERCISES

6.6.1 DISCUSSION QUESTIONS

1. Why would we use agent-based models instead of analytic models?
2. Define two of the agent behavior types.
3. Explain the phenomenon of emergence in SoS. Why is this property important?
4. What is the "ultimate goal" of ABM? Explain what is gained from this "ultimate goal."

6.6.2 EXERCISES

1. Imagine you are a decision-maker about to allocate millions of dollars toward an SoS. The analysts have told you that the SoS configuration they recommend was developed using an agent-based model. What are the top three questions you would ask the analysts before making your decision? Justify each question with 1-2 sentences.
2. Choose one of the items below to complete:
   a. Prepare an abstraction depiction/'paper model' and complete pseudo-code that would implement a fire escape ABM for a university classroom. Pseudo-code requires more than a text description of steps, but less than actual code; a text description and definition of variables/functions used is considered acceptable. Make sure that your pseudo-code covers all aspects of the ABM as presented: objects/agents, space, time, dynamics, etc.
   b. Prepare an abstraction depiction/'paper model' and complete pseudo-code that would implement an ABM that models the interaction of autonomous and human-piloted air vehicles in a notional Urban Air Mobility (UAM) operational context and that is intended to study airspace safety. Pseudo-code requires more than a text description of steps, but less than actual code; a text description and definition of variables/functions used is considered acceptable. Make sure that your pseudo-code covers all aspects of the ABM as presented: objects/agents, space, time, dynamics, etc.

7 Specialized Methods and Tools for System of Systems Engineering

As clarified in part I of this book, Systems-of-Systems present various features that make them a different entity than monolithic systems, however complex they might be. For this reason, not only existing methods and tools need to be modified and adapted to the specific needs of SoS, but also new methods and tools need to be developed, which specifically address features and traits of SoS problems. In this chapter, we provide examples of some of these methods and tools which have been developed *ad hoc* for SoS engineering.

7.1 ANALYTIC WORKBENCH

The Analytic Workbench (AWB) is a collection of methods and techniques that have been developed by researchers at Purdue University starting in 2011. This research project began as part of an effort in SoS engineering initiated by the United States Department of Defense (DoD), and is continuously receiving improvements and extensions based on the application of methods on SoS problems proposed by various stakeholders, including MITRE, NASA, and the US Navy.

Due to the complex and multifaceted nature of SoS modeling and analysis, the most effective approach is to develop different methodologies, each addressing one specific aspect of SoS, for example emergence due to interactions or presence of multiple stakeholders. The AWB implements this approach by presenting the user with a set of tools developed on purpose for modeling and analysis of SoS, and a common representation as a network of interdependent systems.

The research effort that resulted in the implementation of the AWB was based on the wave model (Figure 7.1), identified by Dahmann et al. [41] as a model of evolution of SoS that expands the traditional *Vee* model for systems engineering. The wave model shows hierarchy and time-scale across the steps in the evolution of an SoS. The hierarchy also ranges from the broad, overarching objectives that are strategic in nature (δ-level) to the technical or tactical aspects of individual systems and subsystems (α-level).

The AWB addresses complexities associated with interconnections that exist across physical, functional, and developmental SoS hierarchies. The idea is to support the "top-down integration, bottom-up implementation" paradigm that is part of the wave model, within a quantitative support framework. The workbench for SoS design and development addresses issues of cost, performance, schedule, and risk.

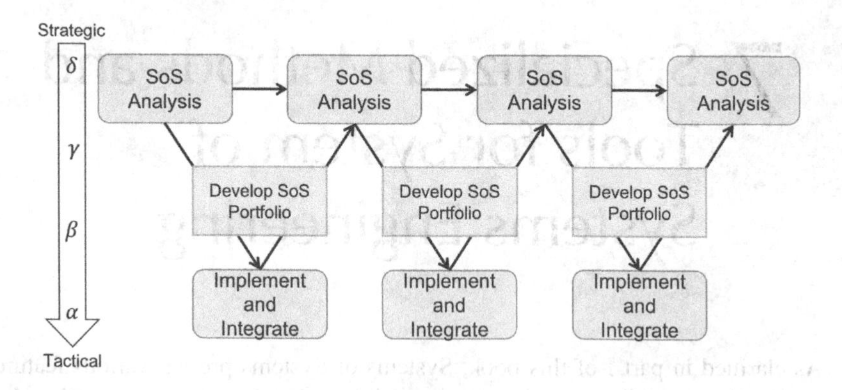

Figure 7.1: SoS wave model.

More specifically, the analytical tools in the workbench account for the complex and highly interconnected nature of the systems that constitute the overall SoS, and allows the user to

1. Quantify performance and risk for individual systems, links and of overall SoS,
2. Assess the impact that changes to SoS architecture (add/remove links and/or nodes) will have, and
3. Quantitatively identify optimal sets of architectural solutions given constraints on cost, performance and risk.

In general, when building tools to support decision-making in an SoS environment, the challenge is that such tools must address the technical and programmatic complexities of SoS, and yet remain domain-agnostic. It is up to researchers to find the appropriate balance between the need for tools that can be used on a broad spectrum of applications in various fields and the need for tools that can be easily tailored to specific applications and user requirements.

Figure 7.2 illustrates the AWB and its primary phases of use in SoS-level analysis and decision-making. The iterative process of the workbench starts with an SoS practitioner's desire to explore a trade-space for analysis and subsequent evolution, based on a target performance/capability metric. The practitioner possesses data that describe the state of the current architecture and also of potential, yet-to-be-introduced systems. The first phase involves the identification of "archetypal questions" that typically arise from the SoS practitioner's technically-motivated queries on assessing the connection between objective metrics and constraints such as cost and schedule. Questions that typically arise include:

- How to assess direct consequences due to changes in architecture?
- Where, what and how much do risks change with operational changes?
- How to mitigate risks?

- Which systems and connections should be added/removed to improve the architecture?

These typical questions reflect the need to examine the coupled behaviors that SoS exhibit, and their consequence on performance/capability metrics. Since the problem is already defined, but we are now characterizing the big picture and assessing the dynamics of systems interaction, this step of the AWB pertains to the abstraction step in the DAI process (Section 3.4). The following phase involves the mapping of these archetypal questions to relevant tool(s) in the workbench. The mapping may suggest employment of multiple tools in the workbench due to overlapping analysis and decision-making requirements. The formal SoS DAI process supports this phase of "choosing the right tool" when transitioning from abstraction to implementation. The iterative process proceeds with the next phase, in which appropriate analysis to address the questions of SoS practitioners is executed in concert with available "truth models" (e.g., computational simulations or field testing), to provide preliminary verification of the next SoS evolution solution. If the user requires only analysis, the tools in the AWB provide different types of useful output and post-processing. If the user requires also synthesis, the solution provided by some of the tools refers to suggested architectural features and architectural changes towards fulfilling target SoS capabilities, while preserving acceptable levels of risk and cost. As the user obtains more data and the architecting of the SoS proceeds, more analysis can be performed as needed, in an iterative refinement process.

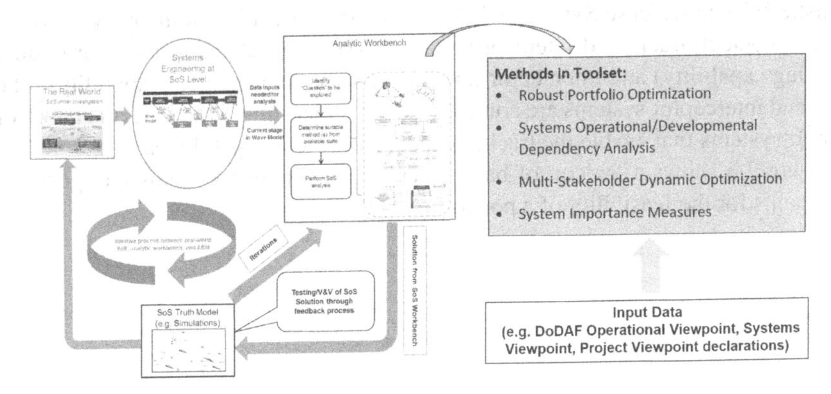

Figure 7.2: Iterative use of the Analytic Workbench (questions, association with tools, design of experiments, analysis of results).

7.1.1 ROBUST PORTFOLIO OPTIMIZATION

7.1.1.1 General Formulation of Investment Portfolio Approach

The problem of modeling and creating an SoS architecture can sometimes be described as a combinatorial problem to find and invest into the most promising portfolio of systems that can achieve a certain desired capability. For instance, in a modular satellite systems architecture, designers need to find the best combination of components, modules, communication systems, and interactions to achieve the goal of the satellite or constellation. The process of selecting the optimal portfolio takes into account the Life Cycle Cost (LCC), the capability of the entire SoS based on the performance of interacting nodes, and the resulting uncertainty or risk resulting from the selected systems.

The Robust Portfolio Optimization (RPO) method seeks to maximize the overall SoS performance measured using a predetermined capability index and to keep both cost and certain types of risk (e.g. developmental, cost, etc.) within acceptable levels while accounting for the impact of various forms of data uncertainty. Using RPO for a particular SoS design problem yields a set of Pareto optimal solutions (each solution is a portfolio of systems) corresponding to a user-defined risk aversion factor. The stakeholders of the complex SoS problem can then explore the design space of available options resulting from the RPO analysis based on dependencies between various systems.

RPO models an SoS as a network of discrete nodes, each with a predefined set of features. Each node has a set of requirements (inputs) and capabilities (outputs). Connection between compatible nodes, based on rules involving the inputs and outputs, allows for feasible architectures to be developed. Each system provides capabilities at the system level. A combination of these system-level capabilities provides SoS-level capabilities. Several systems provide support capabilities necessary for adequate operation of all the systems in the selected portfolio. For example, while the Ballistic Missile Defense System (BMDS) is a conglomeration of many systems, its ability to detect, track, and intercept threats is primarily related to its radar system (tracking capability) and interceptor missiles systems (engagement capability). The radar and interceptor systems are connected to an underlying hierarchical network of support systems that enable the tracking and engagement capabilities.

Figure 7.3 shows a lexicon of basic interactions between systems that generate constraints for the feasibility of a portfolio.

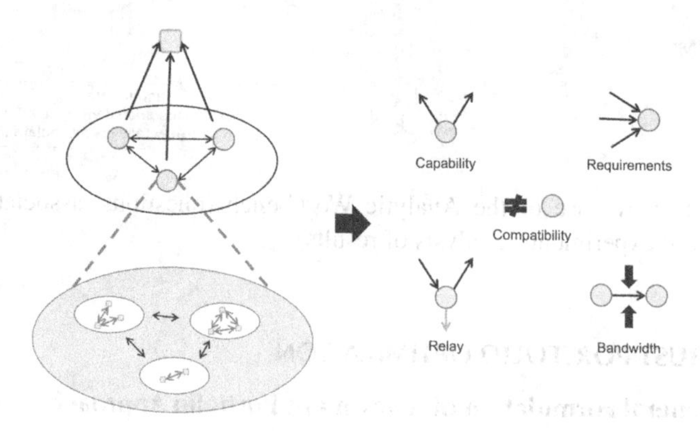

Figure 7.3: Archetypal node (system) behaviors.

These generic behaviors model basic, aggregate interactions between systems as simple nodal constraints and behavior that are applicable to a wide variety of inter-system connections as typically described in systems engineering architectural processes. While not exhaustive, the combinations of these nodal behaviors as modeling rules can cover a large set of real-world inter-system interactions. The five most intuitive nodal interactions are:

1. *Capability*: Nodes have a finite supply of capabilities that are limited by the quantity and number of their connections.
2. *Requirements*: Nodes have requirements to enable inherent capabilities. Requirements are fulfilled by receiving connections from other nodes that possess a capability to fulfill said requirements.
3. *Relay*: Nodes can have the ability to relay capabilities between adjacent nodes. This can include excess input of capabilities that are used to fulfill node requirements.
4. *Bandwidth*: The total amount of capabilities or number of connections between nodes are bounded by the "bandwidth" of the connection linkages between systems.
5. *Compatibility*: Nodes can only connect to other nodes based on a pre-established set of connection rules.

The performance of an SoS is measured by the ability of its connected network of individual systems to fulfill overarching core objectives. SoS-wide performance is quantified by a combination of the capabilities of nodes that most directly contribute to the core objectives. The nodes in this case can also represent small clusters of systems that have a defined set of inputs (requirements) and outputs (capabilities), which determine their aggregate behavior. This generic definition of nodes enables the methodology to be applied to both the system and SoS levels of analysis. The notion of selecting a "basket" of individual systems based on performance, connectivity to other systems, and uncertainty in performance bears a resemblance to that of an investment portfolio problem, thereby allowing use of tools and methods imported from financial engineering.

Figure 7.4 shows the efficiency frontier based on the amount of risk and return for a given portfolio. Such an efficiency frontier enables stakeholders to choose a portfolio that has a maximum return for an accepted risk level.

The generic problem of optimizing a portfolio of systems or technologies based on requirements, constraints, and uncertainty can be developed in different *flavors*, based on the type of uncertainty. The different methods of portfolio optimization are discussed as follows.

7.1.1.2 Robust Mean Variance Optimization

Mean Variance Optimization (MVO) is a well-known financial portfolio theory developed by [124] that provides a way to select a diverse portfolio of assets in a manner that leverages risk against reward. The SoS investment model can be posed as an MVO investment problem. The objective is to maximize the expected network performance in fulfilling key overarching objectives while keeping developmental

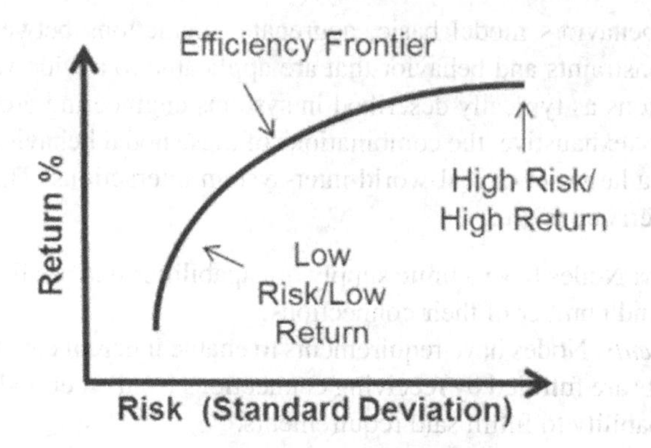

Figure 7.4: Efficiency frontier of optimal portfolios for given investor risk averseness.

risk and cost below acceptable levels. Each candidate system is assumed to have a fixed estimated cost comprising its acquisition cost and relevant integration costs with other systems to achieve the intended function(s). The selection of systems is constrained by the requirement to provide generic connectivity capabilities between constituent systems. To add robustness to MVO, an SDP approach as developed by [161], which addresses uncertainty in the expected returns and covariance matrix. The uncertainty here simply translates to a linear "safety margin" in the context of a system's capability. The mathematical formulation for the SoS investment problem while using Robust Mean Variance Optimization (RMVO) is given by Equations 7.1 — 7.15 and is known as a Mixed Integer Quadratic Program (MIQP). Solving this MIQP results in a set of Pareto optimal portfolios that balance expected rewards against risk (variance).

$$\max \quad \left(\sum_q \left(\frac{S_{qc} - R_c}{R_c} w_c x^B_{q \in nc} \right) - \lambda(\langle \overline{\Lambda}\hat{\Sigma} \rangle - \langle \underline{\Lambda\Sigma} \rangle) \right) \tag{7.1}$$

$$\text{subject to} \quad \sum_j x_{cij} \leq x^B_i S_{ci} \tag{7.2}$$

$$\sum_i x_{cij} \geq x^B_j S_{rj} \tag{7.3}$$

$$x_1 + \ldots + x_n = L \tag{7.4}$$

$$\sum_c x_{cij} - x_{ij} M \leq 0 \tag{7.5}$$

$$M \sum_c x_{cij} - x_{ij} \geq 0 \tag{7.6}$$

$$\sum_j x_{ij} \leq \text{Limit}_i \tag{7.7}$$

$$x^B_j S_{cj} + \sum_i x_{cij} - \sum_j x_{cij} - x^B_j S_{ij} = 0 \tag{7.8}$$

$$x_{cij} \leq \text{Limit}_{cij} \tag{7.9}$$

$$x_{cij} = 0 \; \forall \; c \in \text{capability} \tag{7.10}$$

$$x_{cij} \in \{\mathscr{Z}, \mathscr{R}\}, \; x_j^B \in \{0,1\} \tag{7.11}$$

$$x_q^F = \frac{x_q^B C_q}{\text{Budget}} \tag{7.12}$$

$$\sum_q C_q x_q^B \leq \text{Budget} \tag{7.13}$$

$$x_q^F \in \{\mathscr{Z}, \mathscr{R}\}, \; x_{ij} \in \{0,1\} \tag{7.14}$$

$$\begin{bmatrix} \overline{\Lambda} - \underline{\Lambda} & x_q^F \\ x_q^{F\prime} & 1 \end{bmatrix} \geq 0 \tag{7.15}$$

where w_c is a weighting factor used to weigh the importance of each capability that contributes directly to the objective function; Σ is the covariance matrix; S_{ci} is capability c of system i; S_{rj} is requirement r of system j; w is a constant weighting factor vector of SoS capabilities; λ reflects the investor's ability to tolerate risk; x_i^B is a binary decision variable for selecting system i; R_c is base SoS capability for normalization, x_{cij} is quantity of capability c between system i and j; x_{ij} is the adjacency matrix that indicates connection between systems i and j; M is a constant value corresponding to the Big-M formulation; Limit_i is the stipulated number of connections for system i.

Equation 7.1 is the objective function that seeks to maximize overall SoS capability while minimizing cost and developmental risk (measured here in development time). The first term in the objective function normalizes the capabilities sought by dividing the potential capability of available systems, S_{qc}, by a baseline performance, R_c. The weights, w_c, are assumed to be already ascertained through methods from literature such as *preferential learning* in [172]. An example on selecting objective weights is also provided in [139]. The second term in the objective function, $\lambda(\langle\overline{\Lambda}\hat{\Sigma}\rangle - \langle\underline{\Lambda}\Sigma\rangle)$, specifies the developmental risk as per the robust formulation devised by [161].

Connectivity and resource constraints at feeder nodes are represented by Equations 7.2 – 7.11. Equation 7.2 is the capacity limit at each node where the total of capability type c going from node i to all other nodes j is limited to the total amount of capability from the node if it were selected $(x_i^B S_{ci})$, where S_{ci} is a constant. This ensures that the individual capabilities at each node do not provide more capability to other parts of the SoS network than what is available at the respective node. For example, the selection of a system with communications capabilities should mean that the total bandwidth of data that is being used by other interfacing systems should not exceed the inherent limits of the communications system. Equation 7.3 ensures that requirement conditions are met for each node and permits the availability of excess capability to be present at the node. This is expressed by the summed flow of capability type c from nodes i into node j is greater than the requirement type r if the node were selected $(x_j^B S_{rj})$. Equation 7.4 enforces compatibility constraints in the form of binary logic.

The restrictions for an SoS architecture needs user-defined inputs on acceptable connectivity constraints in the form of the number of connections between nodes. Equations 7.5 – 7.7 are associated with connection restrictions between nodes that follow a "Big-M" formulation approach, which establishes a logic condition that if x_{cij} is non-zero, the corresponding x_{ij} value must also be non-zero (in this case, a value of 1). The Big-M formulation thus allows for tracking connections between systems and permits control of the number of connections to individual systems through the adjacency variable, x_{ij}. The drawback of the Big-M approach is that a poor selection of the penalty term M can result in poor relaxations of the original integer problem. The poor quality of the relaxed problem can thereby lead to computational complexities in finding solutions. This has led to the development of algorithms that address the issues of selecting appropriate values of M, based on the specific problem formulation, such as illustrated by [138].

The summation in Equation 7.7 enforces the maximum allowable number of links for each selected system, expressed as the sum of all incoming connections from systems j to system i, to be less than Limit_i. Equation 7.8 is a flow balance that ensures preservation of "capability flows" about the interconnected network of systems, such as power, connectivity, and bandwidth between systems. This ensures that the amount of capability c *within* a node and the amount of capability c *received* by the same node satisfy the requirements that need a capability c and outflow requirements to other nodes with the same capability. The flow constraint of Equation 7.8 is useful for situations where a node has the ability to use incoming resources and relay the excess resource to other nodes as well. Equation 7.9 provides a constraint on the number of connections a node's capability c can make between node i and node j. Equations 7.11 and 7.14 show that the decision variables can be integer, binary, or real numbers, a choice that depends on individual system properties.

Equation 7.12 is the budget fraction, quantified as the fraction of the total budget that each system represents. Equation 7.13 ensures that the total cost of systems acquired is within the prescribed budget. It is also assumed that the budget used in Equations 7.12 and 7.13 is a known and fixed quantity. Equation 7.15 shows a linear matrix inequality for robust formulation as specified by [161].

A more general treatment of capability gains that reflects diminishing value in having an additional unit of a particular choice (e.g., number of systems of the same type) can be modeled using piece-wise linear representations in the objective function. Additionally, cost and capability of individual systems are related such that a more expensive system of the same type may yield better capability. This can be modeled by treating the different options as separate candidate nodes with a constraint that only one of these can be selected.

Risk is captured in terms of variance and moderated by the risk aversion factor λ that reflects the investor's ability to tolerate risk. The negative sign in the objective function indicates a penalty that attempts to balance the gains from selecting individual systems with the accumulated risks. In the context of an SoS investment framework, the risk refers to project development time of and between interconnected systems. The risk in delays can be related to extra work hours and thus can be quantified as an explicit cost. For new, yet-to-be-designed systems, estimations of the distribution curve for development time are done by design intuition and follow traditional project management guidelines in systems engineering as listed in literature such as [22, 109, 171].

The RMVO formulation also requires that the covariance matrix Σ be positive semi-definite. This is to negate the possibility of having portfolios of assets with zero net portfolio risk, a condition that cannot intuitively exist. However, the arbitrary construction of individual entries of the covariance matrix requires a correction to be made, should the resulting matrix violate the positive semi-definiteness requirement. There are a range of methods that allow for tailoring the covariance matrix to fulfill specific characteristic behaviors. Given a covariance matrix Σ, the resulting "closest" positive semi-definite matrix can be found by minimizing the Frobenius norm between the two as shown by [36].

An example problem for RMVO method is contained in [44]. By solving a problem with the RMVO method, a Pareto or an efficiency frontier of portfolio solutions is generated that shows SoS performance for a given variance in development time of a portfolio (or an SoS solution). Two other variants of RPO incorporating operational and simulated risks are presented as follows.

7.1.1.3 The Bertsimas-Sim Method

The Bertsimas-Sim method is a robust linear formulation that addresses parametric data uncertainty without excessively penalizing the objective function. This method helps obtain desired portfolios with SoS capabilities based on operational risks. This method allows for the control of probability of constraint violations and the effect of the degree of conservatism on the objective function, also known as the price of robustness. Its linear formulation makes it naturally extendable to discrete optimization problems and a very attractive method for application to the current (linear) SoS architectural framework. The formulation starts by addressing the general inequality constraints in a traditional linear programming problem: $Ax \leq b$. A subset of matrix A_{ij} contains uncertain entries (derived from data) that exist within symmetric intervals, i.e., belonging to set J_i and $[a_{ij} - \hat{a}_{ij}, a_{ij} + \hat{a}_{ij}]$. A conservatism parameter, Γ_i, is used to adjust the degree of conservatism in protection of the uncertain linear constraints from infeasibility. Γ_i takes values in the interval $[0, \mid J_i \mid]$ where J_i represents the set of A_{ij} coefficients that are uncertain and is not necessarily integer. Introduction of uncertainty in the coefficients results in a non-linear problem formulation. However, a proof is derived by [18] that converts the non-linear form into the following linear optimization problem:

$$\max \quad \left(\frac{\sum_q S_{qc} - R_c}{R_c} w x_q^B \right) \tag{7.16}$$

$$\text{subject to:} \quad \sum_j A_{ij} x_j + z_i \Gamma_i + \sum_{j \in J_i} p_{ij} \leq b_i \tag{7.17}$$

$$z_i + p_{ij} \geq \hat{a_{ij}} y_j \tag{7.18}$$

$$-y_j \leq x_j \leq y_j \tag{7.19}$$

$$l_j \leq x_q^B \leq y_j \tag{7.20}$$

$$p_{ij}, y_{ij}, z_{ij} \geq 0 \tag{7.21}$$

The linear formulation shown in the above equations preserves sparsity of the original A_{ij} matrix - an attractive feature for computational efficiency. Equation 7.16 is the objective function that maximizes a general linear function. Equations 7.17 – 7.20 are the robust version of linear inequality constraints where Γ_i is the constant that dictates the level of conservatism in the constraint. $\hat{a_{ij}}$ in Equation 7.18 is the uncertainty associated with the j-th entry of the i-th constraint in the A matrix. Like RMVO, the optimization problem in Bertsimas-Sim method is also subject to satisfying a range of connectivity and resource constraints as specified by Equations 7.2 – 7.11.

The uncertainty intervals in Equations 7.16 - 7.21 and 7.2 - 7.11 describe the parametric uncertainty associated with the capability coefficients for each system (node) across the SoS network. The formulation of the Bertsimas-Sim approach, as specified in [18], is utilized to include uncertainties within the optimization problem. A large benefit of using the Bertsimas-Sim approach is the ability to maintain a Mixed Integer Program (MIP) formulation whilst providing the means to select solutions based on probabilistic bounds of constraints violation, without excessively penalizing the objective function as shown in [18]. Each constraint reflects the ability of the SoS architecture to maintain a specific capability. An example problem involving Bertsimas-Sim optimization method is provided in [42].

7.1.1.4 Conditional Value-at-Risk Optimization

A more recent measure of risk, developed by financial engineers at J. P. Morgan, is the Value-at-Risk (VaR) measure that defines percentiles of loss and represents predicted maximum loss with a specified probability level over a defined time horizon as specified in [163]. VaR measure evolved into the Conditional Value-at-Risk (CVaR) that represents a weighted average between the VaR and the losses exceeding the VaR measure; this is important as protections against VaR alone do not limit exposures to the maximum losses that can be incurred should worst-case scenarios be realized. The CVaR formulation to managing portfolio risk is very attractive since it does not require explicit construction of complicated joint distributions in the formulation, results in a linear programming (LP) problem, and satisfies sub-additivity of risks. The CVaR formulation incorporates simulated risks while obtaining portfolios that have desired CVaR values.

A detailed derivation of the linear programming counterpart of CVaR can be found in [60] and [163]. The formulation assumes a linear loss function associated with each asset, as is typically the case for holding financial assets. The SoS portfolio optimization problem is posed as a mathematical programming problem that seeks to minimize the SoS performance index CVaR exposure as quantified from agent simulated operational losses in SoS level performance. The resulting equations are:

$$\min_{x,z,\gamma} \left(\gamma + \frac{1}{(1-\alpha)S} \sum_{s=1}^{S} z_s \right) \tag{7.22}$$

subject to:

$$z_s \geq \sum_i (b_i - y_{is})' x_i^B - \gamma \tag{7.23}$$

$$\sum_i b x_i^B \geq \mathrm{SoS}_{cap} \tag{7.24}$$

$$z_S, x_i \geq 0 \tag{7.25}$$

Equation 7.22 is the objective function that seeks to minimize the CVaR of selecting a collection of SoS level assets that directly contribute to the SoS performance index calculations. This objective function comprises the value at risk term, γ, and weighted summations of the simulated loss scenarios, $s = 1 \ldots S$, at the prescribed confidence level, α. Equation 7.23 is the inequality constraint associated with the loss incurred for each simulated scenario where b_i is the expected return and y_{is} is the stochastically simulated return scenario, s, for asset i; the number of scenarios, S, represents the total number of Monte Carlo simulations run. y_{is} can be simulated via stochastic partial differential equations (SPDEs) or via a black box Artificial Neural Network (ANN). Equation 7.23 thus incorporates agent-based simulated outcomes of potential SoS level performance losses through the vector y_{is}, which represent the simulated outcomes in performance for system i under scenario s. A scenario has the system i deployed alongside other systems in the agent model simulation and under different mission operation scenarios. More specifically, the parameters for key drivers of the mission are changed for each simulation run (e.g., effective range of radar, simulated communication links being compromised due to enemy actions, etc.). The number of scenarios is equal to the number of agent simulation runs required to reasonably approximate the outcomes of the SoS architectures.

Equation 7.24 ensures a minimum SoS level of performance as constrained by the constant, SoS_{cap}. The optimization problem is solved using a range of values to generate the Pareto frontier that trades off SoS level performance for SoS value at risk. The optimization problem is also subject to satisfying a range of physical and operational constraints as specified by Equations 7.2 – 7.11 (same as RMVO and Bersimas-Sim methods). The resulting frontier represents the optimal set of portfolios that best trade off expected return against CVaR. In the context of an SoS development framework, the frontier provides the trade-offs between performance and anticipated worst case scenario losses at the prescribed confidence level. An example problem utilizing CVaR optimization method is available in [43].

7.1.2 SYSTEMS OPERATIONAL DEPENDENCY ANALYSIS

Systems Operational Dependency Analysis (SODA) is a method to model and analyze systems input/output behavior in the operational domain. It is based on a network-like representation of operational dependencies between systems. SODA analysis can compute the *operability* of each system as a function of its own internal status, and of the operability of the other systems in the network, based on the topology and the features of the dependency. The model can therefore evaluate the impact and propagation of failures and disruptions from the nominal operability, and the effect of architectural design decisions on the global behavior. Based on these results, the user can quantify various metrics of interest, for example robustness and resilience. Using SODA, designers and decision-makers can quickly analyze the operational behavior of SoS and evaluate different architectures under several working conditions. The user can compare architectures based on metrics of interest, trade-off between competing desired features, identify the most promising architectures, as well as the causes of the observed behavior, and discard architectures that lack the requested features. This way, in the early design process, promising architectures can be taken into consideration, and improved based on the information given by the model parameters and the observed behavior, thus supporting the process of concept selection.

7.1.2.1 Genesis of the Model

SODA models and analyzes the result of possible cascading effects of dependencies between systems on the overall operability, in case of disruptions. It is based on the Functional Dependency Network Analysis (FDNA) method [69, 70], a two-parameter model of dependencies between capabilities [71]. This method is derived from the input/output model of [117] for infrastructures [91, 92].

FDNA is a two-parameter piece-wise linear model of dependencies between capabilities. The parameters have intuitive meaning, and the method is used to analyze the impact of failures on the desired capabilities in complex systems. SODA retains the idea of a simple parametric model, as well as the concept of separating the impact of a one-to-one dependency into a critical zone, where very low input is available, and a non-critical zone, where high input is available. Extending FDNA, SODA considers the influence of the internal status of a system and adds stochastic behavior modeling. Further enhancements include the modeling of various combinations of multiple inputs.

SODA is a three-parameter piece-wise linear model, suitable for analyzing system dependencies, including partial dependencies, and their effect on system behavior. Features of SoS and the computational cost to perform analysis of these systems require analytical methods to keep some inherently qualitative aspect. However, one of the goals of SODA is to keep these qualitative features to a minimum. For this reason, a SODA parametric model trades a more detailed behavioral analysis for a quantitative representation of the system's behavior. The parameters of SODA model may be evaluated via parametric regression analysis from experimental or historical data but also may be estimated, if such data are not available, by expert assessment. SODA has been validated through ABM simulations and multiple research projects.

7.1.2.2 Operational Dependencies

SODA models the architecture of an SoS as a directed operational network $G = (N, E)$, where N is a set of nodes and E is a set of edges, where each edge is an ordered pair of nodes (Figure 7.5). The nodes represent either the component systems or the capability to be acquired. Accordingly, the edges represent the operational dependencies among the systems or among the capabilities.

In SODA, each node is characterized by its internal health status, or Self-Effectiveness (SE), ranging between 0 (system not working at all) and 100 (system at maximum performance). Each edge is characterized by three parameters: Strength of Dependency (SOD), Criticality of Dependency (COD), and Impact of Dependency (IOD), that affect the behavior of the entire SoS in different ways. These parameters, described in detail in the next subsection, can be determined by expert judgment and evaluation, and they yield a direct insight into the cause of the observed behavior. The model parameters can also be computed based on historical data or through a limited number of simulations and experiments.

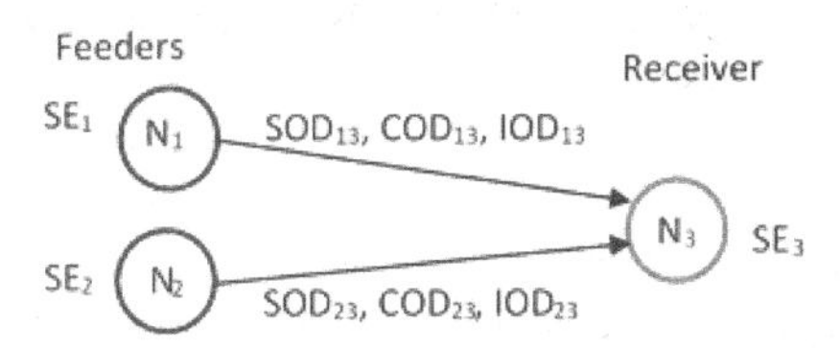

Figure 7.5: Synthetic operational dependency network. N: node; SOD: strength of dependency; COD: criticality of dependency; IOD: impact of dependency; SE: self-effectiveness.

SODA is used to evaluate the effect of topology and of possible degraded functioning of one or more systems on the operability of each system in the network. The analysis can be a deterministic evaluation of the status of a single instance of the system, or a stochastic evaluation of the overall behavior. In the deterministic analysis, given the SE of each system and the properties of each dependency, SODA quantifies the operability O_i of each node according to the model described below. The operability of a node, ranging between 0 and 100, is defined as the level at which the system is currently operating, or the level at which the desired capability is being currently achieved. It is a function of the SE and the dependencies from feeder nodes. The operability of a system is related to performance by means of a given function, as shown in Figure 7.6. It is thus related to the value or utility that the system is achieving; however, a value function may include desired variables other than operability. When designing the operational network, the user must also define the relationship between performance and operability.

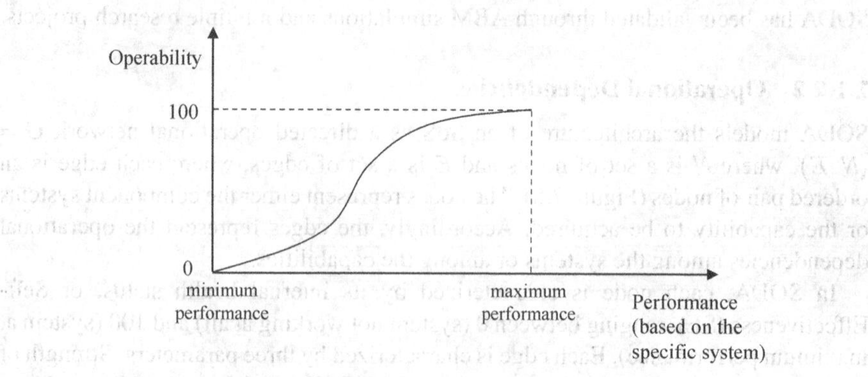

Figure 7.6: Performance and operability. The minimum or worst possible performance corresponds to operability equal to 0. The maximum or best possible performance corresponds to operability equal to 100.

The operability of the nodes of interest is used to analyze and evaluate properties of the overall system, such as robustness, resilience, and risk. In the stochastic version of SODA, the SE of each system follows a probability distribution. Consequently, also the operability of each node is probabilistic. In contrast to the deterministic version, this type of analysis deals with the general behavior of the system, rather than with a single instance. SODA can thus be used to identify the most critical nodes and dependencies in the network under different disrupted conditions in terms of the impact on operability. Designers can compare different architectures and use metrics based on operability to quantify the robustness and resilience of complex systems and the risk of each architectural design in terms of the impact of disruptions. These metrics may be defined based on the specific application and on the desired analysis. Equation 7.35 define metrics for *robustness*, based on the comparison between the retained overall operability and the maximum disruption of any system in the network. If flexibility is added to the architecture by applying rules that allow the architecture to be reshaped in case of disruption, such as allowing a system to support another system that has been disrupted, the comparison between the overall operability when actions are taken to counteract the effect of disruption and the maximum disruption of any system in the network is used as a measure of the *resilience* of the architecture. The *resilience* is defined as the capability to recover part of the loss in operability. This metric is shown in Equation 7.36.

7.1.2.3 Model Parameters

Each dependency between two systems is represented with three parameters. Experiments and simulations on real applications showed that this model is more accurate than other models of the same family, yet still simple enough to guarantee fast analysis of many architectures. The low number of parameters and their intuitive meaning make them suitable both to be assessed by knowledgeable designers and to be used to drive decision-making in the architecture design of complex systems.

Strength of Dependency accounts for how much the operability of a system depends on the operability of a feeder system. SOD is the predominant factor when the feeder has a high level of operability (see SOD zone in Figure 7.7). For each operational dependency, the SOD parameter α_{ij} ranges between 0 and 1, and is defined as the fraction of operability of node j that depends on the operability of node i. The rest of the operability of node j depends on its own SE (internal status). α_{ij} is the slope of the blue line in Figure 7.7. High α_{ij} corresponds to high SOD.

Criticality of Dependency is one of two parameters that quantify how the functionality of a system degrades when a feeder system is experiencing a major failure. COD is the predominant factor when the feeder has a low level of operability (COD zone in Figure 7.7). For each operational dependency, the COD parameter β_{ij} ranges between 0 and 100, and is defined as the maximum loss in operability of node j. This is equal to the drop in the operability of node j when node j has operability equal to 0. In Figure 7.7, β_{ij} is the difference between 100 and the intercept of the 'O_j due to SOD' line with the y-axis. High β_{ij} corresponds to high COD.

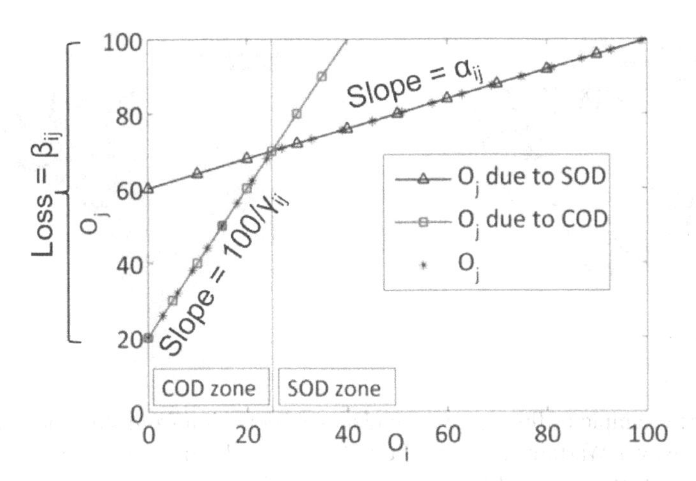

Figure 7.7: Operability due to single dependency of system j from system i. $SE_j = 100$. Line with triangular markers: term due to SOD (here, $\alpha_{ij} = 0.4$). Line with square markers: term due to COD and IOD (here, $\beta_{ij} = 80$, $\gamma_{ij} = 50$). Stars: resulting O_j as a function of O_i. Criticality is prevalent for $O_i < 25$.

Impact of Dependency is an additional parameter that defines the critical zone of the I/O model. In the definition of FDNA [69, 70, 81], the slope of the function relating the operability of nodes i and j in the COD zone is always equal to 1. This means that starting from the minimum operability of node j (low levels of operability of node i) in the COD zone, this operability O_j would depend solely on the operability of node i, without any contributions from the SE of node j. Results from ABM simulations showed that starting from the baseline corresponding to $O_i = 0$, the operability of node j can increase faster than the operability of node i. The critical dependency can thus have a lower impact and be restricted to a smaller zone. A small width of the critical zone models a dependency that may be highly critical, resulting in a large loss of operability if the input is completely disrupted, but that requires just a small amount of operability of the feeder node to achieve high level operability in the receiver node. Both simple linear models and the FDNA piece-wise linear model fail to capture this feature, common in many systems dependencies, resulting in larger modeling error (Figure 7.8). To correct this shortcoming, SODA has an additional parameter γ_{ij} used to quantify IOD. It ranges between 0 (which signifies that IOD is not included in the model) and 100, and is defined as 100 divided by the slope of the COD-dependent function (Figure 7.7). The resulting support function can model a wider spectrum of dependencies than FDNA. For example, it can model dependencies that exhibit an input/output behavior similar to a step function. Other models such as non-linear functions or higher-order polynomials may result in a better behavioral model of complex one-to-one dependencies than SODA, but they will also increase both the computational cost of the analysis and the complexity of the model setup.

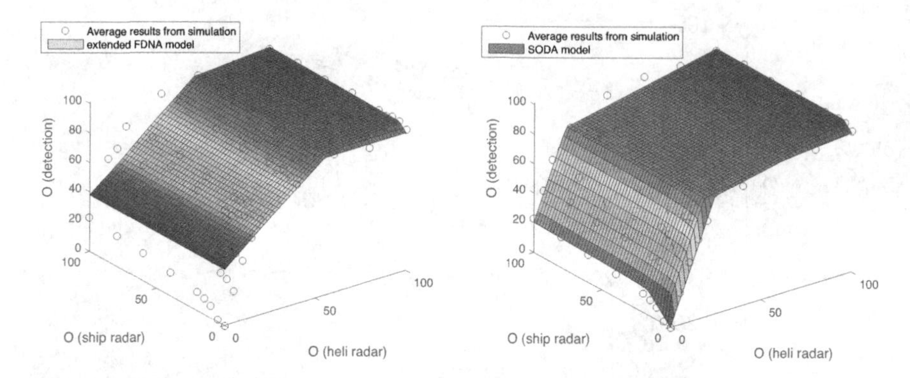

Figure 7.8: Example of fitting the dependency of detection from ship and helicopter radar in a Naval Warfare Scenario. Left: FDNA model does not capture the high increase in detection capability caused by a slight increase in helicopter radar capability and has large errors in the zone where the operability of both feeders is low. Right: SODA model better captures the input/output relationship, and with the corrective weight $\lambda = 0.1$ it better models the combined effect of multiple dependencies.

7.1.2.4 Modeling Dependency on a Single System

Based on the parameters described in the previous sections, SODA models the operability O_j of node j, depending only on node i, according to the following equations.

The operability of root nodes (i.e., nodes that do not depend on any other node) is equal to their SE.

$$O_i = SE_i \tag{7.26}$$

The operability of node j that depends on only onc feeder node i is computed as the minimum of two terms (see Figure 7.7), one depending on the SOD, one depending on the COD and the IOD.

$$O_j = \min\left(O_j^S, O_j^C\right) \tag{7.27}$$

The term depending on the SOD is computed based on the operability of the feeder node i and the SE of the receiver node j.

$$O_j^S = \alpha_{ij} O_i + (1 - \alpha_{ij}) SE_j \tag{7.28}$$

The term depending on the critical zone is computed based on COD, IOD, and the operability of the feeder node i.

$$O_j^C = (100 - \beta_{ij}) + \frac{100}{\gamma_{ij}} O_i \tag{7.29}$$

SODA results in a better model than FDNA due to the addition of the IOD parameter. The root mean square error (RMSE) of SODA modeling is always lower than the corresponding RMSE of FDNA modeling (Figure 7.9).

7.1.2.5 Modeling Dependency on Multiple Systems

When a node depends on more than one feeder node, the equations of SODA are slightly modified. Following the FDNA model of [69, 70], SODA computes the operability term depending on SOD as the average of the corresponding terms for each dependency and the operability term depending on COD and IOD as the minimum of the corresponding terms for each dependency, thus reflecting the intuitive idea of the overall impact of a critical dependency.

However, the use of this formulation results in a possible non-zero operability of a node even when all its feeders have zero operability (Figure 7.8, left). ABM simulations showed that there are cases of dependency from multiple nodes when operability of a node may decrease to zero when all the feeders have zero operability. To retain the simple model of one-to-one dependencies, this effect is modeled by applying a multiplicative weight W to the parameter that models the COD. For each feeder of the node under consideration, the weight is the average of the operability of all other feeders to the same node, and has an exponent λ ranging between 0 and 1. A value of $\lambda = 0.1$ is usually a good compromise to model most dependencies. With the multiplicative weight, the COD parameter β_{ij} represents the loss in operability resulting from the total loss of one feeder when all other feeders are working properly. The exponent models the magnitude of the impact of other feeders in the critical zone. An exponent $\lambda = 0$ models an "OR-like" dependency, where each dependency of node j on node i results in a certain level of operability, without accounting for

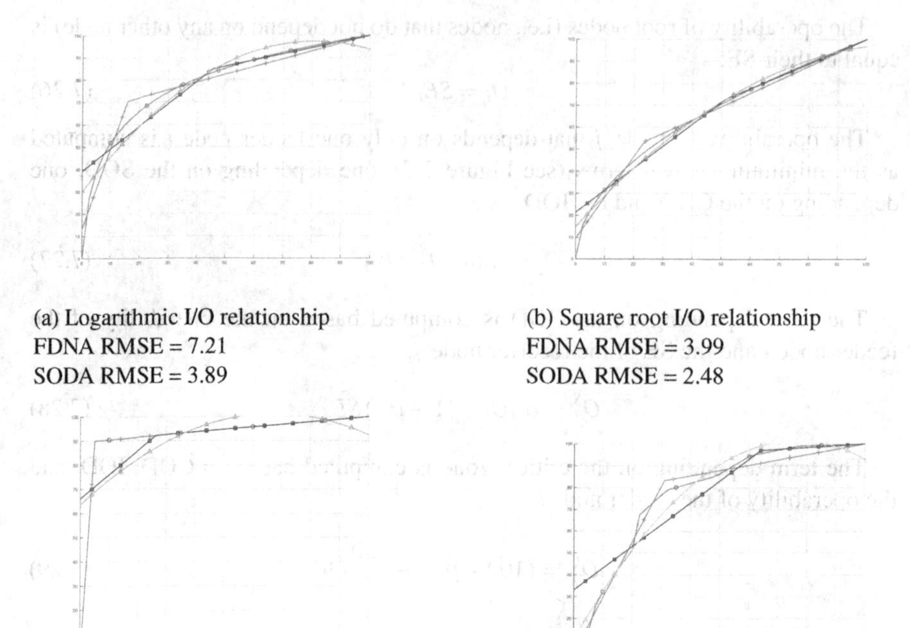

(a) Logarithmic I/O relationship
FDNA RMSE = 7.21
SODA RMSE = 3.89

(b) Square root I/O relationship
FDNA RMSE = 3.99
SODA RMSE = 2.48

(c) Step-like I/O relationship
FDNA RMSE = 2.22
SODA RMSE = 0.02

(d) User-defined I/O relationship
FDNA RMSE = 15.05
SODA RMSE = 1.56

Figure 7.9: Modeling input/output relationships with SODA, FDNA, and polynomials. The I/O relationship is drawn with circles, FDNA model with squares, SODA model with pluses, and degree-3 polynomial model with triangles.

dependencies of node j on other nodes. An exponent $\lambda \neq 0$ models a partially "AND-like" dependency, where the operability resulting from the dependency of node j on node i degrades to 0 if node j is not receiving adequate input from its other feeder nodes. Increasing the exponents models increasing sensitivity on the input of other feeders in the critical zone. Multiple dependencies are then modeled as follows:

The operability of root nodes is equal to their SE:

$$O_i = SE_i \tag{7.30}$$

The operability of a node j that depends on multiple feeders is defined as the minimum of two terms, one depending on the feeder SODs, and one depending on the CODs.

$$O_j = \min\left(O_j^S, O_j^C\right) \tag{7.31}$$

The term depending on the SODs is the average of the SOD-based terms, based on the operability of each of the n feeder nodes and the SE of node j.

$$O_j^S = \frac{1}{n}\sum_{i=1}^{n} O_{ij}^S \tag{7.32}$$

$$O_{ij}^S = \alpha_{ij} O_i + (1 - \alpha_{ij}) SE_j \tag{7.33}$$

The term in the critical zone is the minimum of the terms based on COD, IOD, and the operability of each of the n feeder nodes.

$$O_j^C = \min\left(O_{1j}^C, O_{2j}^C, \ldots, O_{nj}^C\right) \tag{7.34}$$

$$O_{ij}^C = (100 - \beta_{ij}) W_{ij}^{\lambda} + \frac{100}{\gamma_{ij}} O_i$$

7.1.2.6 Evolution of Operability over Time and Robustness

SODA can be used to evaluate the evolution of the operability over time. The results of SODA can be used to build better metrics for robustness of part or all of the life cycle of the SoS. The evolution of the distribution of SE of each system has been evaluated based on a simple failure model. SODA can then quantify the evolution of operability based on the topology of the network and the SE of the systems at each time step. This computation relates the reliability of the individual systems to that of the whole SoS.

7.1.2.7 Robustness and Resilience

Based on these results, the user can quantify the robustness of an architecture, and use it to compare different architectures or to trade off with other measures of performance. Defining robustness as the capability of a system to keep high performance following disruptions, SODA uses the following metric for robustness (with stochastic analysis and continuous time):

$$Rob_{\mathscr{A}} = \frac{1}{n}\sum_{i=1}^{n}\left(\frac{1}{t_f}\int_{t=0}^{t_f}\left(\frac{\frac{1}{m}\sum_{j\in\mathscr{I}} O_j(t,n) - \min_{N\in\mathscr{A}} SE_N(t,n)}{100 - \min_{N\in\mathscr{A}} SE_N(t,n)}\right)\right) \tag{7.35}$$

where $\mathscr{A}$ is an architecture, represented by a network of nodes and edges $G = (N,E)$ and appropriate matrices of SOD, COD, and IOD, n is the number of instances (in SODA stochastic model, using Monte Carlo simulation), t_f is the final time, m is the number of nodes of interest, used to compute the *overall capability* of the architecture, $\mathscr{I}$ is the subset of nodes of interest, and N are all the nodes in architecture $\mathscr{A}$. This formulation requires the lowest SE in the network to be strictly less than 100, to avoid having a 0 value in the denominator of the term in the integral.

At each time instant, the difference between the average operability of interest and the lowest SE in the network is compared to the difference between 100 and the lowest SE in the network. Thus, this is the fraction of maximum loss in a node that is recovered in the average operability of interest (thanks to the operational dependencies). A robustness value of 1 means that the operability of interest is not affected by the loss, and the architecture has maximum robustness. A robustness value of 0 means that the highest disruption affects the overall operability without any recovery based on the topology of the dependencies. This would happen, for example, if the node experiencing the disruption has no dependency from any other node (so that its operability is equal to its SE) and all the paths in the network from the node experiencing the disruption to the nodes of interest have maximum strength (so that the operability of every node following the disrupted node is depending only on this disrupted input). These values of robustness at each time instant are averaged over time (through an integration) and averaged over the stochastic instances. In case of discrete time steps, the integral is replaced by a sum.

It is important to underline a few considerations about this metric:

1. There are many possible metrics for robustness that can be built based on the results of SODA in terms of operability of the systems. The user can decide to tailor this metric to a specific problem or even use a completely different metric.
2. This metric is a *relative* measure of robustness, meaning that it is computed using the difference between the average operability and the minimum SE. The reason is that robustness is not meant to measure the absolute performance of the network, but how well the network is able to absorb disruptions. Intuitively, this means that a topology having an average operability of interest equal to 80 when the minimum SE in any node is 40 is more robust than a topology having the same average operability of interest when the minimum SE in any node is 70.
3. For the same reason, this metric is a ratio, i.e., the percentage of loss that is recovered. Intuitively, this means that a topology "recovering" an amount of operability equal to 20 when the maximum loss in any node is 20 is more robust than a topology exhibiting the same amount of "recovery" when the maximum loss in any node is 80.
4. While this metric is meant to be a function of the topology only, the simple practical formulation also depends on the evolution of the SE of each node over time, and on the interval of integration.
5. This metric has the objective of facilitating architecture comparison, not of identifying the criticalities. Therefore, it does not distinguish which node is experiencing the maximum loss in SE. Other metrics can be easily formulated which consider losses only in one node, so as to measure the robustness to failures on a specific system.
6. For the same reason, this metric does not distinguish between the case of single failure and multiple failures. Once again, more metrics that will distinguish robustness in case of single failure from robustness in case of an unlimited number of failures can be easily derived from this baseline metric.

Thanks to flexibility, an architecture can exhibit resilience. Resilience has many different definitions in literature [162], but it is mostly described as the capability to react to failures and degradation in order to recover operability, at least partially. Applying the same concepts that led to the formulation of a quantitative metric for the robustness of an architecture, SODA utilizes the following formulation of a metric for resilience of an architecture:

$$Res_{\mathscr{A}*} = \frac{1}{n}\sum_{i=1}^{n}\left(\frac{1}{t_f}\int_{t=0}^{t_f}\left(\frac{\frac{1}{m}\sum_{j\in\mathscr{I}} O*_j(t,n) - \min_{N\in\mathscr{A}} SE_N(t,n)}{100 - \min_{N\in\mathscr{A}} SE_N(t,n)}\right)\right) \tag{7.36}$$

where $\mathscr{A}$ is the baseline architecture, represented by a network of nodes and edges $G = (N,E)$ and appropriate matrices of SOD, COD, and IOD, n is the number of instances, t_f is the final time, m is the number of nodes of interest, used to compute the *overall capability* of the architecture, $\mathscr{I}$ is the subset of nodes of interest and N are all the nodes in architecture $\mathscr{A}$. In this case, $O*_j$ indicates the operability of node j in the baseline architecture $\mathscr{A}$ if the conditions are nominal, and in the alternative architecture $\mathscr{A}*$ if the conditions require to switch to the alternative architecture. This formulation requires the lowest SE in the network to be strictly less than 100, to avoid having a 0 value in the denominator of the term in the integral.

At each time instant, the difference between the average operability of interest and the lowest SE in the network is compared to the difference between 100 and the lowest SE in the network. Different from the metrics for robustness, the operability of interest can be higher due to the flexibility of the architecture. Therefore, the metric evaluates the fraction of maximum loss in a node that is recovered in the average operability thanks to both the operational dependencies and the flexibility of the architecture. A resilience value of 1 means that the operability of interest is fully recovered following the loss, and the architecture has maximum resilience. A resilience value of 0 means that the highest disruption affects the overall operability without any recovery due to the architecture or the flexibility. These values of resilience at each time instant are averaged over time (through an integration) and averaged over the stochastic instances. In case of discrete time steps, the integral is replaced by a sum.

Similarly to the considerations relative to robustness, there are a few important points about this resilience metric:

1. There are many possible metrics for resilience that can be built based on the results of SODA in terms of operability of the systems, on possible flexibility or on stakeholder decisions. The user can decide to tailor this metric to a specific problem, or even to use a completely different metric.
2. This metric, similar to the metric for robustness, is a *relative* measure of resilience, meaning that it is computed using the difference between the average operability with flexibility and the minimum SE in the baseline architecture. The reason is that resilience is not meant to measure the absolute performance of the network, but how well the network is able to recover after disruptions.

3. For the same reason, this metric is a ratio, i.e., the percentage of loss that is recovered.
4. This metric is a function of the topology and flexibility. Different rules of available flexibility, including redundancy, added and removed systems and edges result in general in different values of resilience. For this reason, the metric is specific to a given alternative architecture $\mathscr{A}*$. SODA and the resilience metric can be used together with existing approaches, to add considerations about complexity, ease of change, and cost of change, for example the methodologies described by Tamaskar et al. [156], Lafleur [115], Beesemyer et al. [16] and Davendralingam and DeLaurentis [44].
5. This simple practical formulation of the metric is also depending on the evolution of the SE of each node over time, and on the interval of integration.
6. The alternative architecture comes into play only when needed, so the user must set also requirements for the onset of the alternative architecture, for example a minimum acceptable threshold of capability of interest from the baseline architecture. This means that, for instances and time intervals when the baseline architecture is in use, the value of resilience is equal to that of robustness. However, when the alternative architecture is in use, the capability of interest is expected to be higher than with the baseline architecture, resulting in an average resilience higher than the average robustness.
7. This metric has the objective of facilitating architecture comparison and evaluation of the available flexibility, not to identify the criticalities. Therefore, it does not distinguish which node is experiencing the maximum loss in SE. Other metrics can be easily formulated which consider losses only in one node, so as to measure the resilience to failures on a specific system.
8. For the same reason, this metric does not distinguish between the case of single failure and multiple failures. Once again, more metrics which will distinguish resilience in case of single failure from resilience in case of an unlimited number of failures can be easily derived from this baseline metric.

7.1.2.8 Deterministic Analysis

The simplest analysis that can be performed with the SODA method is a single-point, deterministic analysis. Values for the SE of each system (their possible degraded statuses) are fed into the equations to compute the actual operability of each system. This kind of analysis can be thought of as a way to answer "what-if" questions. For example, if the user is interested in the impact of a specific system on the overall behavior of the whole SoS, values ranging between 0 and 100 can be assigned to the SE of the system under consideration, and the subsequent operability of each of the other systems can be computed through SODA and listed in tables or plotted in graphic format. Another example may involve partial failures in several systems, which allow the user to evaluate the combined effect of multiple failures. Chapter 10 shows examples of graphical representations of this analysis.

Deterministic analysis gives good insights into the influence of dependencies on the operational network. The user can identify the most critical nodes under specified conditions; that is, the nodes that most affect the operability of other nodes. Some of the results may show unexpected behavior, such as possible reduction of the impact of failures, due to the particular topology of the network. This kind of behavior is the result of emergence in SoS, and cannot be easily predicted by knowledge of the behavior of individual systems alone. The user can compare different architectures based on their response to failures and accidents. Once critical nodes have been identified, the user can assess the cause of the observed results based on the parameters of the dependencies, and thus evaluate possible countermeasures to maintain an adequate level of operability. It must be noted that the results of this analysis also depend on the output of interest: for example, a node might be critical to the operability of a low-interest node, while having a small impact on a node of interest. Compared to deterministic analysis, stochastic analysis better identifies details about the overall impact of disruptions on the operability.

7.1.2.9 Stochastic Analysis

A more realistic understanding of a system's behavior as a function of the dependencies among components can be achieved using a stochastic SODA analysis. In stochastic analysis, the SE is defined using a probability density function (PDF) rather than a constant value. The corresponding output is a probability density function for the operability of each of the component systems, accounting for all the SOD, COD, and IOD values as well as the overall effect of the system topology. This behavior is condensed in a few probability distributions of interest. In particular, the expected value of the operability of a system gives a measure of the robustness of such system to failures of the feeders, while the variance of the operability represents the sensitivity of the system to failures in the feeder nodes. Such outputs show behavioral patterns and features of a whole architecture and thus are valuable to design activities. For example, given the expected distribution of SE of the component systems over time (including aging, minor failures, and major accidents) and a threshold for the minimum operability to be achieved by some systems of interest, the user can compute the probability that these systems are operating above the given threshold over time. The user can then compare alternative architectures, identify their critical systems, and explain the role of topology in the observed criticality.

An analytic expression for the expected value and variance of the operability can be derived from the SODA equations. In this formulation, operability is a piece-wise linear function of the SE. However, due to the possible high number of nodes in the network, and since SODA is relatively computationally inexpensive (Table 7.1), Monte Carlo simulation appears to be the best choice to perform this type of analysis. Stochastic analysis, similarly to deterministic analysis, can be used to analyze the combined effect of multiple failures.

Since the SODA formulation for multiple dependencies does not explicitly address the correlation between systems feeding the same node, the PDFs of the SE of these systems are statistically independent in the model. Correlation can be addressed in two ways. The first alternative is for the user to choose input for the Monte

Carlo simulation from statistically dependent PDFs. A more accurate alternative is to model the correlation between the systems: since the internal status of the systems is correlated, part of their operability depends on a common cause, which can be explicitly modeled in SODA by a node feeding the correlated systems.

Table 7.1
Computational cost of SODA analysis. Time to perform analysis of a single instance of the operational network, with variable number of nodes and edges. Processor Intel Core i3-2350M 2.3 Ghz, 4Gb DDR3 RAM.

Number of nodes	Average number of edges (100 runs)	Average time	Maximum time
100	2484	0.0125 s	0.0143 s
500	62361	0.887 s	0.922 s
1000	249687	6.625 s	6.804 s
2000	999785	61.41 s	67.82 s

7.1.2.10 Synthesis and Architectural Design Updates

While quantitative SODA analysis allows for comparison between various architectural designs and analysis of the trade-off between competing features, the intuitiveness of the parameters used in the SODA model can support decision-making in design updates. Since the parameters give insights into the causes of the observed behavior, they also suggest possible ways to effectively improve this behavior within an existing architecture and allow for evaluation of the impact of architectural changes. For example, the user can assess the outcome of additional robustness or redundancy of critical nodes, usually characterized by weak SE and high impact on dependent nodes.

7.1.2.11 SODA Problem Setup

Since SODA relies on abstraction and generalization, it can be applied to various problems in different fields. However, this capability also entails the need for the user to understand the principles of this methodology in order to be able to correctly apply SODA analysis to specific problems. The methodology is applied in five steps.

1. **Operational Dependencies.** The first step necessary to apply SODA is the conversion of the requirements and the systems involved into a network of operational dependencies. The topology of this network will depend on the desired or required topology of interactions among the systems (exchange of information, matter, energy), and on the desired output from the analysis:

Table 7.2
Example of Self-Effectiveness and Operability for a Naval Warfare Scenario

SoS	System	Self-Effectiveness	Operability	Metrics
Naval Warfare Scenario	Ship Recon Systems	Probability of the radar to detect adversary within range	Time to detect adversary	Performance, robustness, resilience (flexibility)
	Helicopter Recon Systems	Probability of the radar to detect adversary within range	Time to detect adversary	
	Ship Weapon System	Probability to engage adversary, when within range	Time to engage the adversary	

the network of operational dependencies will include nodes representing the systems that are assigned to perform a task, and nodes representing capabilities that the user is interested to quantify. SODA is a model in the operational domain; therefore, the flow between systems is generalized in terms of operability. Systems engineering methods for functional allocation can be used to perform this step [105]. Functional modeling method [102, 154] has the advantage of identifying possible failure modes [153, 160] that can also be used as a basis for analysis of the impact of disruptions with SODA. Alternative architectures may be characterized by different systems allocated to perform the required function, different network topology, or different features of the dependencies.

2. **Self-Effectiveness and Operability.** The second step in the problem setup is the definition of the internal status, which will be represented by the SE of each systems, and of the performance, which will be represented by the operability of each node. Both SE and operability are normalized between 0 (worst case) and 100 (best case). Multi-dimensional performance is represented by different capability nodes (and associated systems). The measure of internal status and performance, to be associated respectively with SE and operability, can be evaluated by experts, or computed from historical data, experiments, or simulations. Tables 7.2, 7.3, and 7.4 show examples of self-effectiveness and operability for specific applications.
3. **Dependency Parameters α_{ij}, β_{ij}, γ_{ij}.** At this point, to be able to model the overall behavior through SODA, the user needs to determine the parameters of each dependency. FDNA formulation [69, 70] describes a way to evaluate α_{ij} as the fraction of operability of a node depending on its feeder

Table 7.3
On-orbit Satellite Servicing Example

SoS	System	Self-Effectiveness	Operability	Metrics
	Operational satellite component	Internal operational status	Overall status due to internal status and inputs	
On-orbit satellite servicing	Operational satellite	Internal operational status	Overall status due to internal status and inputs	Dynamic performance over time, robustness to aging and failures, resilience due to servicing
	Servicing satellite		(impacts self-effectiveness of serviced components)	

Table 7.4
Example of Self-Effectiveness and Operability for Cybersecurity

SoS	System	Self-Effectiveness	Operability	Metrics
IT SoS	IT systems	Internal status: generation and transmission of correct information	Generation and transmission of correct information, given internal status and possible cyberattacks	Performance, robustness, resilience (flexibility)

while accounting for a virtual baseline operability of the system when the feeders are not giving any input and there is no criticality. It also suggests to evaluate β_{ij} as the operability that a system achieves when the feeders are not giving any input. The parameters for the dependencies can be effectively evaluated by experts based on these definitions and on the equations of SODA if the users have a good understanding of the impact of each of these parameters on the operability dependency between a feeder node and a receiver node. Figure 7.10 shows the effect of α_{ij} on a single dependency of node j on node i when $\beta_{ij} = 95$, $\gamma_{ij} = 10$, and α_{ij} increases from the left plot to the right plot. Each plot in the figure represents the operability O_j as a function of the operability O_i. As the SOD increases, the operability of node j is increasingly dependent on that of node i and less dependent on its own SE.

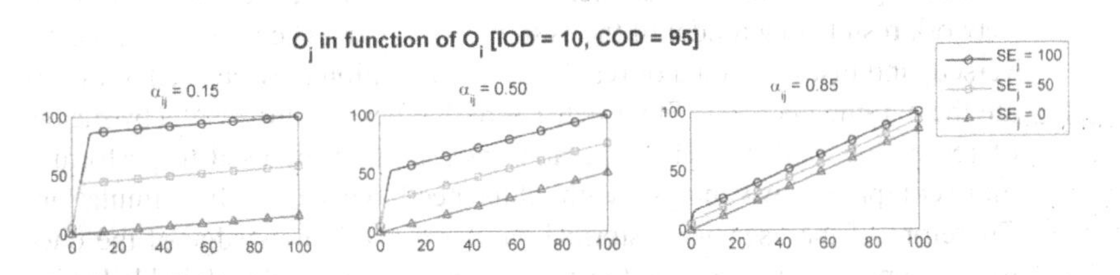

Figure 7.10: Effect of SOD on operability dependency of node j from node i. SOD increases from left to right.

Figure 7.11 shows the effect of β_{ij} on a single dependency of node j on node i when $\alpha_{ij} = 0.5$, $\gamma_{ij} = 40$, and β_{ij} increases from the left plot to the right plot. Each plot in the figure represents the operability O_j as a function of the operability O_i. As the COD increases, the critical zone (low values of O_i) expands and the loss due to criticality increases.

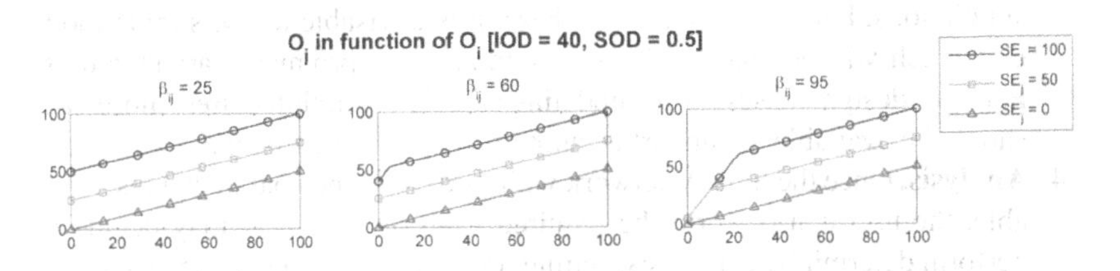

Figure 7.11: Effect of COD on operability dependency of node j from node i. COD increases from left to right.

Figure 7.12 shows the effect of γ_{ij} on a single dependency of node j on node i when $\alpha_{ij} = 0.5$, $\beta_{ij} = 95$, and γ_{ij} increases from the left plot to the right plot. Each plot in the figure represents the operability O_j as a function of the operability O_i. As the IOD increases, the critical zone (low values of

O_i) expands, having impact even at high values of O_i, especially when SE_j is high.

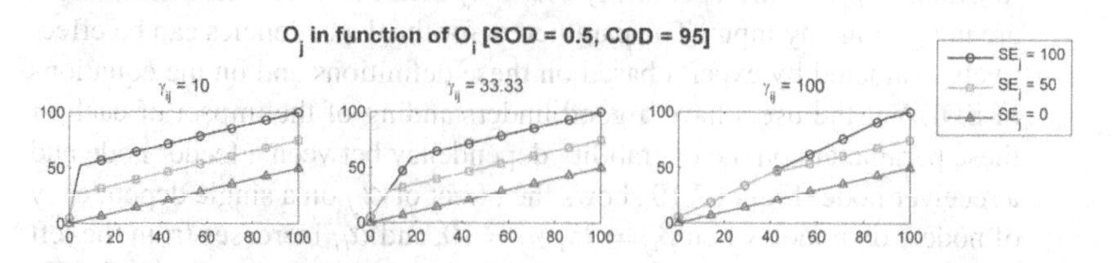

Figure 7.12: Effect of IOD on operability dependency of node j from node i. IOD increases from left to right.

Combining these three parameters with multiple dependencies within the network results in a model of the system's behavior. The process can be reversed, and historical data or results from simulation may be used to evaluate the parameters of the SODA structure that best fits the results by means of regression analysis. These parameters may then be used to perform a more comprehensive analysis without the need to execute a full simulation. Therefore, if the user has historical data or a good knowledge of the one-to-one dependency behavior, the parameters that model the global behavior can be computed with a cost of the same order of magnitude as the cost of other parametric models such as response surface methodology. This cost is generally lower than the cost of performing full, high-fidelity simulations. If data are not available, Design of Experiments techniques can be applied so that a low number of simulations can be used to identify the parameters of the model. Since SODA is a parametric model, it trades detail for low cost and for the advantage of having a representation capable of giving insight into the reason for the observed results. The level of detail produced by a SODA analysis will depend on the level of detail of available data or simulations. For the early design phase, it is advisable to use simple models, which will be able to identify more and less promising architectures. For later design phases, or for updating an existing architectures, more data should be available to support accurate parametric regression.

4. **Analysis.** Once the SODA network with its dependency parameters is available, the user can perform the required analysis. The user may decide to perform deterministic analysis, setting up points of interest (each one being a set of SE values for each node, corresponding to specific failures), and use the results to evaluate the impact of different failures, including multiple failures, and to identify the critical systems and dependencies. Otherwise, the user may use models of failure probability in individual systems to perform stochastic analysis to determine behavior of the whole entity and the main effects of failures and functional dependencies. Metrics of interest representing, for example, the robustness and resilience of the entire system, can be built based on the results of deterministic or stochastic analysis

and their evolution over time.

5. **Synthesis.** SODA allows for fast analysis of the effect of various failure scenarios on systems behavior and capabilities, as well as comparison of architectures and possible trade-offs between competing metrics such as reliability, resilience, capabilities, and cost. The parameters of the SODA model for operational dependencies give insight not only into the effect of the operational dependencies among the systems, but also into some of the reasons why an observed behavior occurs. Based on the relationship between some of the parameters, the topology of the network, and the observed results, the designer can determine the appropriate actions needed to improve the architecture. These actions may include increasing redundancy, adding or deleting paths between node pairs, and investing money to increase the robustness of critical nodes. New architectures can be generated based on these observations, and re-analyzed by means of SODA. This process can be repeated to improve the architecture of an SoS.

7.1.2.12 Source of Parameters

7.1.2.12.1 Design of Experiments and Data Fitting.

Analogously to the approach followed for other surrogate models such as response surface methodology, the first possible source of parameters is based on Design of Experiments (DoE) and data fitting. Once the user has defined the meaning of self-effectiveness and operability for their specific problem, data points can be generated at various levels of operability and then fitted using a SODA model. The choice of the data points can be more or less computationally expensive, ranging from Full Factorial Design [63] (Figure 7.8) to Fractional Factorial Design [30, 129] and the array method formulated by Taguchi [155]. In general, a higher number of points will result in a more accurate model. Data points may be obtained from experiments, simulations, and/or historical data.

Given a network of n nodes $G = (N, E)$, each of the k data points is characterized by the self-effectiveness level of each node. SE_i^a is the self-effectiveness of node i at data point a. For each data point, experiments, simulation, or archived results give the corresponding data-based operability level of each node, D_i^a. Given a SODA model characterized by parameters α_{ij}, β_{ij}, and γ_{ij} for each edge $(i, j) \in E$, the same self-effectiveness will result in the corresponding model-based operability level of each node O_i^a. At this point, the user can perform a parametric fitting, minimizing the square error between the data points and the outcome of the SODA model. This problem is unconstrained but bounded.

$$\min_{\alpha_{ij},\beta_{ij},\gamma_{ij}} \sum_{a=1}^{k} ||[O_1^a(\alpha,\beta,\gamma)\cdots O_n^a(\alpha,\beta,\gamma)] - [D_1^2\cdots D_n^a]||^2 \tag{7.37}$$

$$\begin{aligned} \text{subject to} \quad & 0 \le \alpha_{ij} \le 1, && (i,j) \in E, \\ & 0 \le \beta_{ij} \le 100, && (i,j) \in E, \\ & 0 < \gamma_{ij} \le 100, && (i,j) \in E. \end{aligned}$$

7.1.2.12.2 Simple Input/Output Models.

The data fitting based on Design of Experiments uses data points generated from the whole system architecture, including multiple dependencies and chains of dependencies. An alternative data-based approach to assessing SODA parameters is to use data resulting from simple input/output models relative to single dependencies between two nodes i and j. This is effective especially at a lower level of abstraction, down to the subsystem and component level. In this approach, the data fitting to a SODA model will be performed on one-to-one dependencies rather than on the whole network. Each set of data will result in a SODA dependency model of the type showed in Figure 7.7. The parametric fitting is again an optimization problem, minimizing the square error between the data points and the outcome of a one-to-one SODA dependency.

7.1.2.12.3 One-to-One Dependency Fitting.

A similar approach to the simple input/output model is based on expert judgment. Since the model fitting is related to only a one-to-one dependency, a user familiar with the specific input/output behavior of the dependency between node i and j can place the points that will be used in lieu of data from experiments or simulations in the parametric regression procedure. The regression will follow Equation 7.37. Figure 7.13(a) shows an interactive MATLAB™ figure that allows the user to input data points for a one-to-one dependency. A MATLAB script normalizes the values so that the highest input corresponds to $O_j = 100$ and performs regression to generate a SODA model. The resulting piece-wise linear model is shown in Figure 7.13(b).

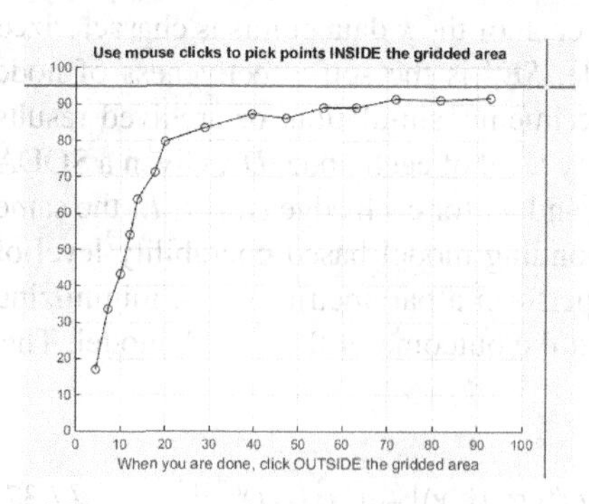

(a) User-input points corresponding to behavioral dependency of node j on node i

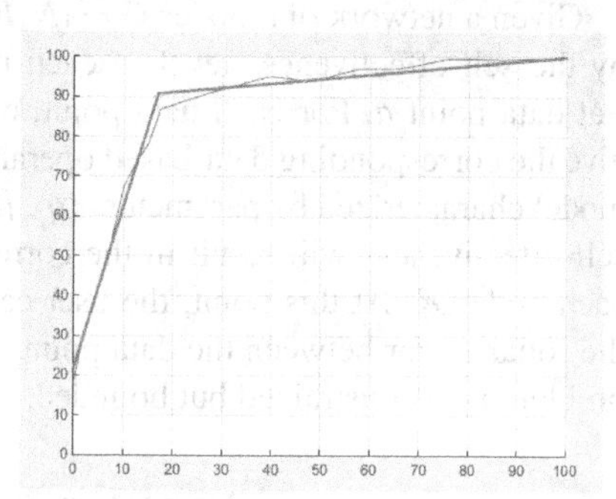

(b) Fitting of SODA model to the input points, after normalization of the maximum value to 100

Figure 7.13: SODA fitting of data input by expert user for a one-to-one dependency.

7.1.2.12.4 Parameter Selection.

Alternatively, based on the intuitive meaning of the three parameters of SODA model for a one-to-one dependency, an expert user can directly input appropriate parameters into the model. Figures 7.10, 7.11, and 7.12 can facilitate this decision by showing the impact of changes in the parameters of the model.

7.1.3 EXAMPLE OF APPLICATION OF SODA: AN EARTH OBSERVATION SYSTEM

This section illustrates the application of SODA modeling and analysis to a simple space problem. The Earth observation SoS shown in Figure 7.14 is composed of two control centers CC_1 and CC_2, one communication satellite (ComSat), a constellation of three observation satellites S_1, S_2, and S_3, and a node representing the sensing capabilities that the user wants to achieve. Since this is a small, simple application meant to illustrate the use of SODA, details about the orbits and visibility of the satellites have not been considered. The network is representative of the operations of the SoS during conditions in which the satellites are in view or can communicate with the control centers using a relay.

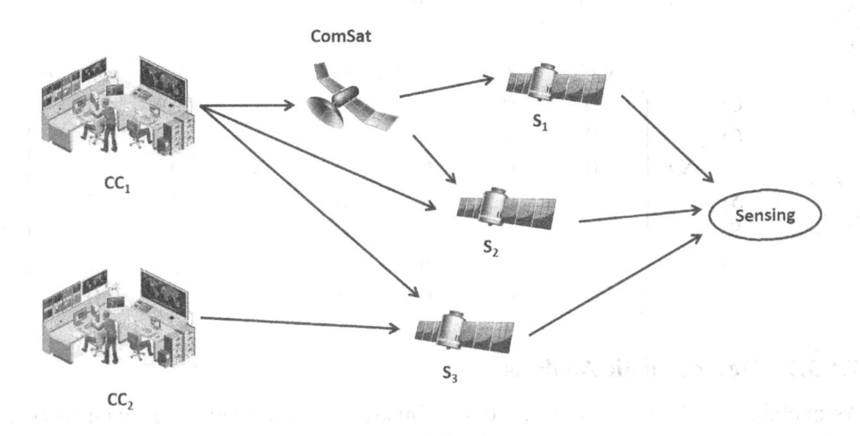

Figure 7.14: Operability dependencies of the Earth observation SoS. CC: Control Center; ComSat: Communication Satellite; S: Observation Satellite.

The values of SOD, COD, and IOD are as follows:

$$
SOD = \begin{array}{c} \\ CC_1 \\ CC_2 \\ ComSat \\ S_1 \\ S_2 \\ S_3 \\ sensing \end{array}
\begin{array}{c} \begin{array}{ccccccc} CC_1 & CC_2 & ComSat & S_1 & S_2 & S_3 & sensing \end{array} \\
\left[\begin{array}{ccccccc}
0 & 0 & 0.3 & 0 & 0.4 & 0.3 & 0 \\
0 & 0 & 0 & 0 & 0 & 0.7 & 0 \\
0 & 0 & 0 & 0.6 & 0.2 & 0 & 0 \\
0 & 0 & 0 & 0 & 0 & 0 & 1 \\
0 & 0 & 0 & 0 & 0 & 0 & 1 \\
0 & 0 & 0 & 0 & 0 & 0 & 1 \\
0 & 0 & 0 & 0 & 0 & 0 & 0
\end{array}\right] \end{array}
$$

$$
COD = \begin{array}{c} \\ CC_1 \\ CC_2 \\ ComSat \\ S_1 \\ S_2 \\ S_3 \\ sensing \end{array}
\begin{array}{c} \begin{array}{ccccccc} CC_1 & CC_2 & ComSat & S_1 & S_2 & S_3 & sensing \end{array} \\
\left[\begin{array}{ccccccc}
0 & 0 & 70 & 0 & 60 & 60 & 0 \\
0 & 0 & 0 & 0 & 0 & 60 & 0 \\
0 & 0 & 0 & 90 & 70 & 0 & 0 \\
0 & 0 & 0 & 0 & 0 & 0 & 40 \\
0 & 0 & 0 & 0 & 0 & 0 & 40 \\
0 & 0 & 0 & 0 & 0 & 0 & 40 \\
0 & 0 & 0 & 0 & 0 & 0 & 0
\end{array}\right] \end{array}
$$

$$
IOD = \begin{array}{c} \\ CC_1 \\ CC_2 \\ ComSat \\ S_1 \\ S_2 \\ S_3 \\ sensing \end{array}
\begin{array}{c} \begin{array}{ccccccc} CC_1 & CC_2 & ComSat & S_1 & S_2 & S_3 & sensing \end{array} \\
\left[\begin{array}{ccccccc}
0 & 0 & 100 & 0 & 50 & 50 & 0 \\
0 & 0 & 0 & 0 & 0 & 10 & 0 \\
0 & 0 & 0 & 25 & 5 & 0 & 0 \\
0 & 0 & 0 & 0 & 0 & 0 & 30 \\
0 & 0 & 0 & 0 & 0 & 0 & 30 \\
0 & 0 & 0 & 0 & 0 & 0 & 30 \\
0 & 0 & 0 & 0 & 0 & 0 & 0
\end{array}\right] \end{array}
$$

7.1.3.1 Deterministic Analysis

As explained in Subsection 7.1.2, deterministic analysis assumes a given *working point* (set of SEs for each system in the SoS), with the goal of assessing the effects of disruptions and reduced operability on the SoS. In this example, each *working point* represents a partial failure and consequent degraded SE in only one system. Table 7.5 shows the corresponding operability level of each sensing satellite and of the overall sensing capability of the SoS given different levels of degraded operability of one node.

These results show that the first control center CC_1 is the most critical system in the architecture. In addition, the overall capability is more sensitive to a partial failure in the second observation satellite S_2 than in the other two observation satellites. Information on criticality can be used both when deciding actions to be taken against the criticality of an individual systems (assuring that the critical system maintains high reliability) and when architecting the entire SoS (such as deciding the number of satellites required to keep the operability above a given threshold in case of failures).

Table 7.5
Deterministic Analysis of the Operability of the Sensing Satellites and of the Overall Sensing Capability of the Earth Observation SoS

Disrupted system	SE	O_{S_1}	O_{S_2}	O_{S_3}	$O_{sensing}$
CC_1	70	94.6	93.1	95.5	94.4
CC_1	25	73	80.5	88.75	80.75
CC_2	70	100	100	89.5	96.5
CC_2	25	100	100	73.75	91.25
ComSat	25	68.5	94.75	100	87.75
S_1	25	70	100	100	90
S_2	25	100	47.5	100	82.5
S_1	25	100	100	62.5	87.5

7.1.3.2 Stochastic Analysis

As an example of stochastic analysis, Figure 7.15 shows the operability of sensing capability following degradation in the control centers and the communication satellite that exhibit SE following a $\beta(2,8)$ distribution. In this case, three systems are experiencing a reduced level of SE at the same time, according to the given probability density function. The expected value of the SE of the control centers and the communication satellite is 20, while the resulting operability of the sensing capability has an expected value of 58. This result gives a first insight into the robustness of the architecture, that is, its ability to maintain a high level of operability following disruptions and partial failures.

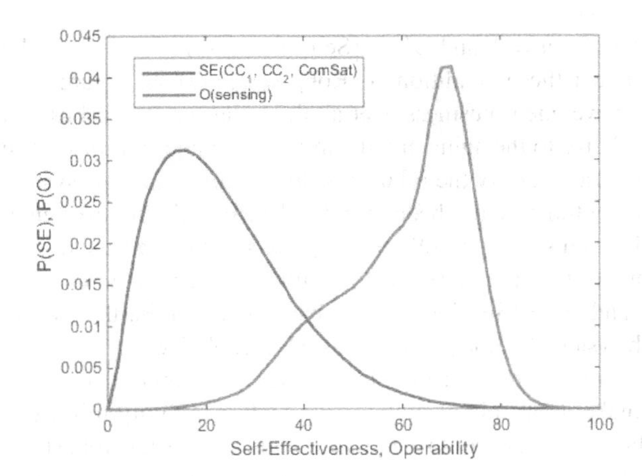

Figure 7.15: Operability of sensing capability of the Earth observation SoS (curve skewed to the right) following disruptions in the control centers and the communication satellite (curve skewed to the left).

In this example, 100 discrete time steps have been simulated and analyzed. Beginning with $SE = 100$, at each time step the control centers, communication satellite, and observation satellites experience a decrease in SE due to aging equal to $0.01 + 0.04\mathscr{U}[0,1]$, i.e., a probabilistic value between 0.01 and 0.05. Furthermore, at each time step each system could experience a minor failure, with a probability of 1% and a loss in SE equal to $3 + 3\mathscr{U}[0,1]$, and a major failure, with a probability of 0.6% and a loss in SE equal to $30 + 20\mathscr{U}[0,1]$. Figure 7.16 shows the evolution over time of the SE of the control centers, communication satellite, and observation satellites in one instance of their life cycle of 100 time steps.

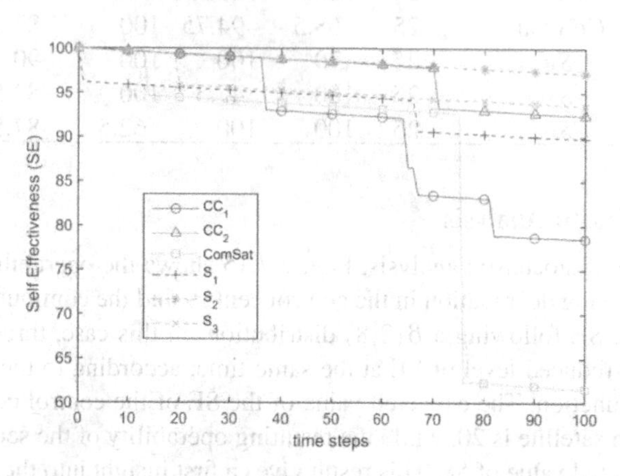

Figure 7.16: Evolution over time of systems SE in the Earth observation SoS. CC_1 experiences multiple minor failures, ComSat experiences a minor and a major failure, and S_2 experiences only decrease in SE due to aging.

With $n = 40000$, $m = 1$ and $\mathscr{I} = \{\text{Sensing}\}$, the robustness of the Earth observation SoS under these conditions is $\text{Rob}_{\text{EarthObs}} = 0.60$. Figure 7.17 shows the histogram of the average robustness over all the instances. The robustness measured in this case is relative to the aging and the minor and major failures modeled.

The user can also quantify the robustness to failures in a single system. A $\beta(8,2)$ probability density function has been used for the SE of the control centers, communication satellite, and sensing satellites (only one system at a time has SE according to this function, while the others have maximum SE), $n = 20000$ per each failed system, $m = 1$ and $\mathscr{I} = \{\text{Sensing}\}$. The robustness of the Earth observation SoS to failures in each system N, $\text{Rob}^N_{\text{EarthObs}}$, is listed in Table 7.6.

The architecture is very robust to single failures, meaning that the redundancies and topology make it capable of maintaining high operability following failures in one system. The results also confirm that the architecture is less robust in case of failures in CC_1 and S_2, analogously to what was observed using deterministic analysis. The average robustness to single failures is $\text{Rob}_{\text{EarthObs}} = 0.82$.

Figure 7.18 shows the robustness over all the instances for single failures. The robustness measured in this case is relative to single failure, and disrupted SE modeled

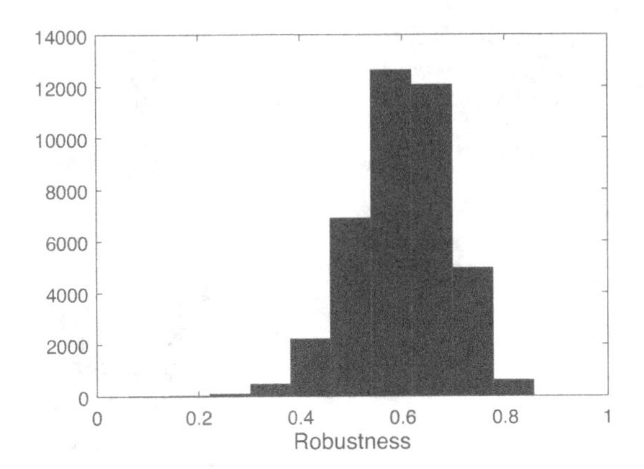

Figure 7.17: Histogram of robustness of the Earth observation SoS with aging and failures over 40,000 instances.

Table 7.6
Robustness of the Earth Observation SoS to Failures in a Single System Modeled with SE $= \beta(8,2)$

Disrupted System	Robustness to Disrupted System
CC_1	0.745
CC_2	0.882
ComSat	0.837
S_1	0.867
S_2	0.767
S_3	0.833

with a $\beta(8,2)$ PDF.

7.1.3.3 Flexibility and Resilience

The architecture can include some flexibility which allows for measuring resilience. For example, suppose that the control centers can perform part of each other's tasks, in the case of degradation of the overall behavior. This means that two new edges become part of the architecture, connecting the CC_2 to Comsat and to S_2. The criticality of CC_1 will be lower in this scenario. Even if this simulation, using the same aging model used to compute robustness, does not consider which is the disrupted system, and just "blindly" applies the alternative architecture when the sensing capability becomes lower than 70, there is still a slight improvement brought by the alternative architecture, with new resilience equal to 0.61 and average sensing capability increasing from 83.6 to 84.8. Figure 7.19 shows the average resilience over all the instances.

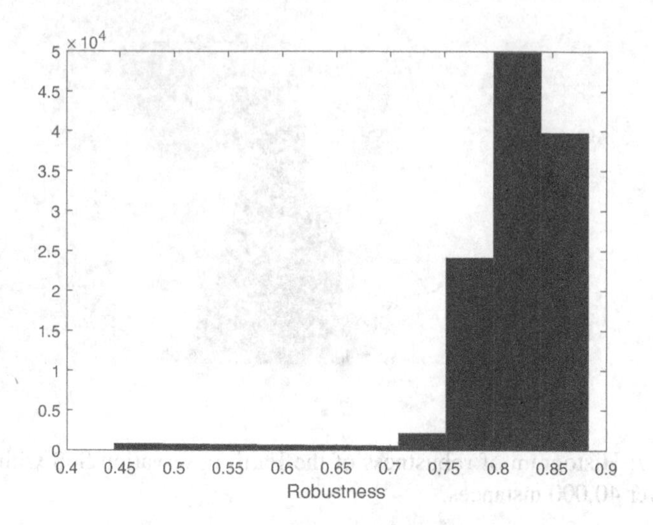

Figure 7.18: Histogram of robustness of the Earth observation SoS with single failures modeled with $SE = \beta(8,2)$ over 120000 instances.

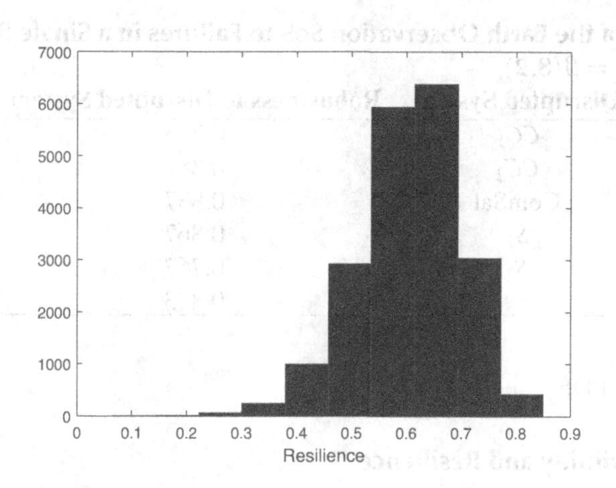

Figure 7.19: Histogram of resilience of the Earth observation SoS with aging and failures over 20,000 instances.

If the alternative architecture is implemented only in case of disruptions in the first control center, the resulting resilience is 0.63 and the average sensing capability is 85.2.

The user can also quantify the resilience to failures in a single system. This analysis yields the most interesting results. Since the alternative architecture is supporting the first control center with the second one, a relatively large increase from the robustness of architecture $\mathscr{A}$ to the resilience obtained thanks to the alternative architecture $\mathscr{A}^*$ should occur when a single failure disrupts the first control center. A $\beta(8,2)$ density function has been used for the SE of the disrupted system, $n = 20000$, $m = 1$, and $\mathscr{I} = \{\text{Sensing}\}$, and an accepted threshold of the operability of sensing equal to 85. Below this level, the alternative architecture is activated. The resilience $\text{Res}^N_{\text{EarthObs}}$ of the Earth observation SoS to failures in each system N is listed in Table 7.7.

Table 7.7
Robustness and Resilience of the Earth Observation SoS to Failures in a Single System (SE Modeled with $\beta(8,2)$)

Disrupted system N	$\text{Rob}^{\text{N}}_{\text{EarthObs}}$	$\text{Res}^{\text{N}}_{\text{EarthObs}}$
CC_1	0.745	0.841
CC_2	0.882	0.878
ComSat	0.837	0.839
S_1	0.867	0.867
S_2	0.767	0.773
S_1	0.833	0.830

This architecture with the given rules of flexibility exhibit high resilience to single failures, meaning that the high robustness is supported by the small changes in the topology resulting in the alternative architecture, and the SoS is capable to recover operability following failures. At first, one might assume that the alternative architecture is generally preferable to the baseline architecture. This is not always the case, however. For example, CC_2 must sacrifice some of its capability in order to support CC_1 in the alternative architecture. For this reason, in case of failure in CC_2, the alternative architecture is less capable to recover operability than the baseline architecture. Table 7.7 shows how the alternative architecture performs much better in case of failures in the most critical systems (CC_1 and S_2), which supports the strategy of switching from the baseline architecture to the alternative architecture in case these kinds of failures occur. The average resilience to single failures is $\text{Res}_{\text{EarthObs}} = 0.84$ when using the alternative architecture in any case where the operability of sensing is below the threshold.

Figure 7.20 shows the resilience over all the instances. When the alternative architecture comes into play, the SoS is more capable to recover from disruptions, resulting in a value of resilience higher than the corresponding robustness.

7.1.4 SYSTEMS DEVELOPMENTAL DEPENDENCY ANALYSIS

Systems Developmental Dependency Analysis (SDDA) is a method used to assess the impact of partial developmental dependencies between components in an SoS. A parametric model of developmental dependencies outputs a schedule of the development of the SoS, accounting for partial dependencies. Using SDDA, the expected *lead times* and the beginning and completion time of the development of each

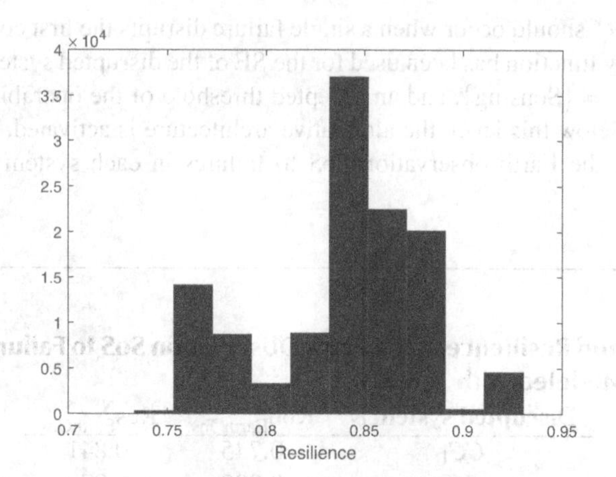

Figure 7.20: Histogram of resilience of the Earth observation SoS with single failures (SE modeled with $\beta(8,2)$) over 40,000 instances.

system can be automatically scheduled and re-scheduled. SDDA evaluates these times based on the current and expected performance of each system (in terms of development time), the model of the dependencies, and the amount of accepted risk. Rules that dynamically change the parameters of the dependencies based on stakeholder decisions or development deadlines can be added to the basic model for more in-depth analysis. SDDA supports educated decision-making in the development and revision phases of systems architecture. In particular, throughout the whole development phase, the information produced by SDDA can be used to identify criticalities and bottlenecks, to quantify possible partial delay absorption, and to assess the best time to begin the development of each system, accounting for development cost, stakeholder decisions, and risk.

Similar to SODA, SDDA borrows the concepts of Strength of Dependency (SOD) and Criticality of Dependency (COD) from the Functional Dependency Network Analysis (FDNA) method, a two-parameter model of dependencies between capabilities. Likewise, SDDA models the developmental dependencies between systems with two parameters, and has an intuitive meaning that facilitates the modeling process.

The outcome of SDDA analysis is the beginning time and the completion time of the development of each system, as well as an assessment of the combined effect of multiple dependencies and possible delays in the development of predecessors. The *lead time*, i.e., the time by which a system can begin to be developed before a predecessor is fully developed, is calculated based on the parameters of the dependencies and the performance of the predecessors. SDDA allows for deterministic or stochastic analysis. When using deterministic analysis, SDDA evaluates the impact of a single instance (i.e., one given amount of delay in each system), resulting in one beginning time and completion time for each system. When using stochastic analysis, the amount of delay in each system follows a given probability density function. Consequently, the beginning and completion time of each system will also be a probability density function. SDDA evaluates the most critical nodes and dependencies with respect to the development time and propagation of delays. Results from the analysis are used to compare different architectures in terms of development time, ability to absorb delays, and flexibility.

7.1.4.1 Developmental Dependencies

The method models the interactions and the exchange of information required for SoS development in the form of developmental networks. These are directed networks $G = (N, E)$, where N is a set of nodes and E is a set of edges, where each edge is an ordered pair of nodes. The nodes represent systems to be developed. As in PERT networks, the links represent developmental dependencies between systems (Figure 7.21). A dependency of a system j on a system i means that system j needs some input such as information or other deliverables related to the development of system i to complete its own development. In contrast with PERT, the dependencies are not absolute and account for partial independence of development of each system. This partial independence is usually accounted for by assigning a fixed *lead time* based on expert judgment. SDDA gives a simple yet more realistic model of this kind of dependency.

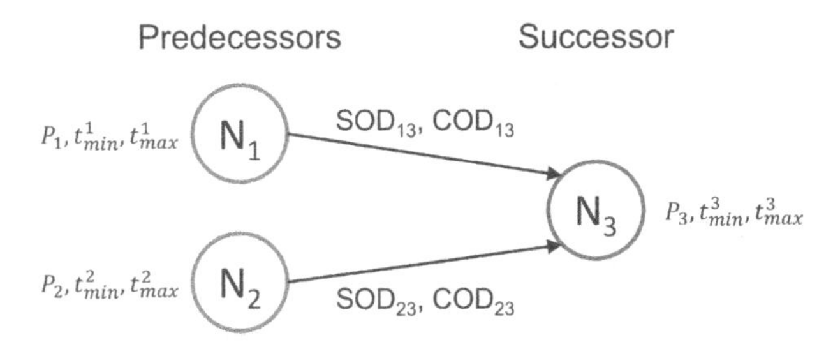

Figure 7.21: Synthetic developmental dependency network. N: node; SOD: strength of dependency; COD: criticality of dependency; P: punctuality; t: development time.

In SDDA, each system i requires three pieces of input data. The first two inputs are the minimum independent development time t^i_{min}, and the maximum independent development time t^i_{max}. These are the minimum and maximum development times of system i, respectively, not accounting for dependencies. The third input is a reliability state, representing the system's timeliness or punctuality P_i. The punctuality, normalized between 0 and 100, constitutes an assessment of how close the system is to being developed in its expected minimum time, not accounting for dependencies. It is analogous to SE in SODA. The higher the punctuality, the more a system is following the expected schedule, thus resulting in a shorter development time. The relationship between time and punctuality might not be linear. For example, 5 weeks could be the shortest time to develop a system, corresponding to a punctuality of 100, and 12 weeks could be the longest time to develop the same system, corresponding to a punctuality of 0, but 10 weeks could correspond to a *satisfaction* of 50%, and therefore a punctuality equal to 50. The user can opt for a simple linear correspondence between development time and punctuality (with high punctuality corresponding to short time, and vice versa, as shown in Figure 7.22). In deterministic analysis, the punctuality of each system will be a single number, while in stochastic analysis the punctuality will follow a probability density function.

Each link (each dependency) requires two parameters, Strength of Dependency (SOD) and Criticality of Dependency (COD), that affect the development schedule of the whole system in different ways. These parameters, described in detail in the next section, can come from expert judgment and evaluation or can be computed based on historical data. The framework to overlap product development activities proposed in [114] suggests a model based on the required exchange of information for parallel development. This approach gives important insights into developmental dependencies, and it can be used as a base for SDDA parameter evaluation.

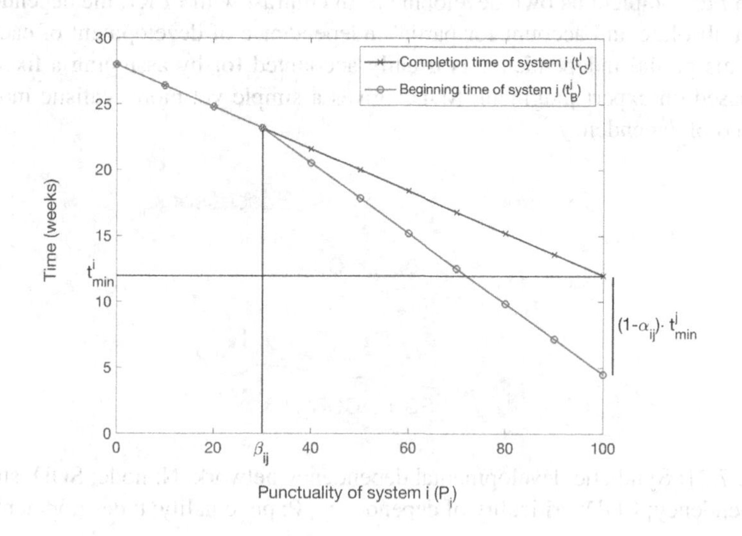

Figure 7.22: Completion time of system i and beginning time of system j as a function of the parameters of the developmental dependency between the two systems ($\alpha_{ij} = 0.25$, $\beta_{ij} = 30$). Due to partial dependency (SOD lower than 1), system j can begin its development before the completion of system i, unless the latter is critically late.

7.1.4.2 Parameters of the Model

Each developmental dependency between two systems is modeled with two parameters. The low number of parameters and their intuitive meaning make them suitable to be assessed by knowledgeable designers and managers, or based upon considerations on information exchange. At the same time, they overcome the inability exhibited by PERT/CPM to manage partial dependencies and dynamic lead times. Figure 7.22 shows the relation between the completion time of a predecessor system i (which is a function of its punctuality P_i) and the beginning time of a successor system j.

7.1.4.2.1 *Strength of Dependency*

The SOD evaluates how much the development of a system is dependent upon input from its predecessor. For each developmental dependency, the parameter for SOD α_{ij} ranges between 0 and 1 and is defined as the fraction of the development time of system i that depends on inputs from its predecessor system j. As shown in Figure 7.22, a system can begin its development before the completion of its predecessor. When the predecessor is developed in its shortest development time ($P_i = 100$), the amount of *lead time* of the successor is equal to its own minimum development time, multiplied by a factor of $1 - \alpha_{ij}$. This means that while the predecessor completes its development, the successor will be able to complete the fraction of its development that does not depend on its predecessor. The value of α_{ij} will trade off between the risk associated with the decision to begin the development early and the possibility of partially absorbing delays thanks to this lead time. A delayed development of the predecessor will affect the beginning time of the successor in two ways. First, the delay will directly be added to the expected beginning time of the successor, like in PERT/CPM methodology. Second, the lead time computed by SDDA will decrease proportionally to the decrease in the predecessor's punctuality (the development of the predecessor is considered to be less reliable), until the punctuality reaches a critical level, under which the lead time is equal to 0 and the dependent node will wait for the full completion of the development of the predecessor.

7.1.4.2.2 *Criticality of Dependency*

The COD is the critical level under which the lead time is equal to 0. For each developmental dependency, the parameter for COD β_{ij} ranges between 0 and 100, and is defined as the normalized punctuality of system i under which system j cannot begin its development before system i is fully developed. Independently from SOD, the COD defines the amount of risk that the manager is willing to take by applying a lead time to the development of the systems. A high criticality means that even a small delay in the development of the predecessor will sharply decrease the lead time, or wipe it out completely.

7.1.4.3 **Basic Formulation of SDDA**

Based on the parameters described in the previous section, on minimum and maximum development time, and on the punctuality of each system, the output of the model is the beginning time t_B^i and the completion time t_C^i of the development of each system i. For a root node (node without any predecessor) i, the beginning time is defined to be 0.

$$t_B^i = 0 \tag{7.38}$$

The completion time, depending on the punctuality, is

$$t_C^i = t_{\text{min}}^i + \left(1 - \frac{P_i}{100}\right)\left(t_{\text{max}}^i - t_{\text{min}}^i\right) \tag{7.39}$$

For a node j having at least one predecessor, SDDA first computes the time necessary for its development t_D^j, based on its punctuality.

$$t_D^j = t_{\text{min}}^j + \left(1 - \frac{P_j}{100}\right)\left(t_{\text{max}}^j - t_{\text{min}}^j\right) \tag{7.40}$$

We then calculate the beginning and completion times based on each dependency from a system i. These are the *actual* beginning and completion times of system j, if it depends on only one system i.

If $P_i < \beta_{ij}$, system i has a critical delay (left side of Figure 7.22). Therefore, the beginning time of system j based on its dependency on system i is equal to the completion time of system i.

$${}^i t_B^j = t_C^i \tag{7.41}$$

Otherwise, the beginning time of system j based on its dependency on i is computed as

$${}^i t_B^j = t_C^i - t_{\text{min}}^j \left(1 - \alpha_{ij}\right) \frac{P_i - \beta_{ij}}{100 - \beta_{ij}} \tag{7.42}$$

In Equation 7.42, the term that is subtracted from the completion time of system i is the lead time of system j. In this basic formulation of SDDA, the actual beginning time of a system j that has more than one dependency is the average of the beginning times resulting from each dependency. This prevents a single predecessor from critically influencing the beginning time.

$$t_B^j = \frac{1}{n}\sum_{k=1}^{n} {}^k t_B^j \tag{7.43}$$

The completion time of system j based on its dependency on system i is

$${}^i t_C^j = \max\left(t_B^j + t_D^j, t_C^i + \alpha_{ij} t_{\text{min}}^j\right) \tag{7.44}$$

where $t_B^j + t_D^j$ is the sum of beginning time and development time, or the completion time that system j would have without accounting for the dependencies. However, the strength of each dependency not only affects the lead time, but it also gives a measure of the fraction of the development time of system j that depends on the full development time of system i. The term $t_C^i + \alpha_{ij} t_{\text{min}}^j$ accounts for this dependency factor, stating that system j cannot be completed before a certain amount of time elapses after the completion of system i.

The actual completion time of system j is the maximum of the completion times given by each dependency.

$$t_C^j = \max_n {}^n t_C^j \tag{7.45}$$

Computation of the beginning and completion times for each node results in a complete schedule of the development of the complex system or SoS, showing the effect of partial development dependencies on the development time.

7.1.4.4 Conservative Formulation of SDDA

In the basic model of SDDA, the actual beginning time of development of a system j is computed as the average of the beginning times resulting from each dependency that system j has on other systems. If the completion time of the predecessors of j spans a large range, this choice could result in an excessive amount of lead time, with consequent increase in cost. To avoid this effect, a more conservative formulation can be used. In this model, called *SDDAmax*, the beginning time is the maximum of the beginning times resulting from each dependency.

$$t_B^j = \max_n {}^n t_B^j \tag{7.46}$$

In the conservative model, Equation 7.46 is used instead of Equation 7.43.

7.1.4.5 Deterministic Analysis

When using deterministic analysis, SDDA evaluates a single instance of the developmental dependencies. Given the parameters of the dependencies and minimum and maximum development times, a single value of punctuality of each system is used to compute the resulting beginning and completion times of the development of each system.

This kind of analysis is useful if the user is interested in the impact of the timeliness of a specific system on the overall development of the complex system. In this case, the user may assign values ranging between 0 and 100 to the punctuality of the system under consideration. The user may then assess the resulting impact on schedule through the analysis of beginning and completion times, listed in tables or plotted in graphic format. Another example may involve delays in several systems, to evaluate the combined effect of multiple postponements.

With deterministic analysis, the user can identify the most critical nodes under specified conditions, that is, the nodes that most affect the development schedule. The user can compare different architectures based on their response to delays. It must be noted that the results of this analysis also depend on the output of interest: for example, a manager might be interested in intermediate deadlines, completion time of development of intermediate systems, partial development of certain systems, and delay absorption in case of low reliability. In the SDDA model, delays can be partially absorbed even on the critical path, due to the partial developmental independence of the systems. Stochastic analysis better characterizes the overall impact of delays on schedule and on the risk associated with unreliable systems in terms of punctuality.

Figure 7.23 shows a Gantt chart [67] for the simple network from Figure 7.21, comparing results from SDDA, SDDAmax, and PERT when all nodes have punctuality equal to 100. The matrices of strength and criticalities of the dependencies are

$$SOD = \begin{bmatrix} 0 & 0 & 0.6 \\ 0 & 0 & 0.75 \\ 0 & 0 & 0 \end{bmatrix} \qquad COD = \begin{bmatrix} 0 & 0 & 25 \\ 0 & 0 & 40 \\ 0 & 0 & 0 \end{bmatrix}$$

The minimum and maximum development times are, in weeks:

$$t_{\min} = \begin{bmatrix} 10 \\ 12 \\ 10 \end{bmatrix} \qquad t_{\max} = \begin{bmatrix} 16 \\ 15 \\ 14 \end{bmatrix}$$

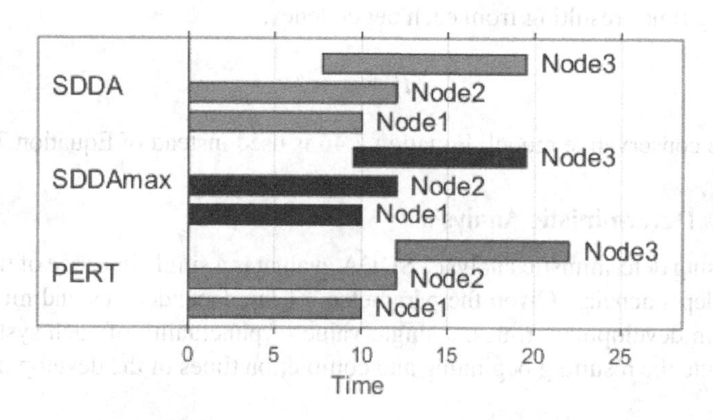

Figure 7.23: Gantt chart showing the schedule of development of the simple three-node network shown in Figure 7.21, according to the basic SDDA, SDDAmax, and PERT models.

Figure 7.23 shows that in PERT analysis node 3 must wait until its two predecessors are fully developed. SDDA and SDDAmax exhibit a lead time for the development of node 3. Due to the high strength and criticality of the dependency of node 3 on node 2, the lead time is small. SDDAmax is more conservative than the basic SDDA model concerning the risk associated with early development of node 3.

Table 7.8 shows the results of various instances of delays in the simple three-node network from Figure 7.21. In this case, the user is interested in the completion time of the whole complex system, that is, the completion time of node 3.

Table 7.8
Results of SDDA Analysis of a Simple 3-Node Network

Punctuality of systems	Model	t_C^3 (weeks)	Maximum delay in single systems (weeks)	Overall delay with respect to baseline (weeks)
	SDDA	19.5	0	0
[100 100 100]	SDDAmax	19.5	0	0
	PERT	22	0	0
	SDDA	19.5	1.5	0
[75 100 100]	SDDAmax	19.5	1.5	0
	PERT	22	1.5	0
	SDDA	22	4.5	2.5
[25 100 100]	SDDAmax	24.5	4.5	5
	PERT	24.5	4.5	2.5
	SDDA	20.25	1.6	0.75
[100 75 100]	SDDAmax	21.29	0.75	1.79
	PERT	22.75	0.75	0.75
	SDDA	21.75	2.25	2.25
[100 25 100]	SDDAmax	24.25	2.25	4.75
	PERT	24.25	2.25	2.25
	SDDA	19.5	1.75	0
[100 100 75]	SDDAmax	20.5	1	1
	PERT	23	1	1
	SDDA	20.75	3	1.25
[100 100 25]	SDDAmax	22.5	3	3
	PERT	25	3	3
	SDDA	22.38	3	2.88
[50 50 100]	SDDAmax	23.08	3	3.58
	PERT	23.5	3	1.5
	SDDA	22.58	3	3.08
[50 100 50]	SDDAmax	23.67	3	4.17
	PERT	25	3	3
	SDDA	21.54	2	2.04
[100 50 50]	SDDAmax	25.08	2	5.58
	PERT	25.5	2	3.5
	SDDA	24.38	3	4.88
[50 50 50]	SDDAmax	25.08	3	5.58
	PERT	25.5	3	3.5

The results listed in Table 7.8 give some interesting insight:

- Under the assumption of partial dependencies, a schedule that follows the SDDA model allows for partial or total delay recovery, and results in a completion time that is lower or equal to that given by a PERT model. This possible delay recovery must be traded against the cost of longer development times of individual systems, and the increased risk due to early decision.
- A schedule that follows the SDDAmax model is more conservative than SDDA, yielding less delay recovery, but also less risk of wasting resources due to early beginning of development. The development of Node 3 is shorter in SDDAmax than in SDDA model, but the overall development time is longer.
- Node 2 is the most critical if it experiences a short delay (e.g., when punctuality is equal to [100 75 100], the delays are higher than other instances with single small decrease in punctuality). Node 1 is the most critical if it experiences a long delay (e.g., when punctuality is equal to [25 100 100], the delays are higher than other instances with a single large decrease in punctuality). Delays in Node 1 are also the most critical when coupled with delays in other nodes.

7.1.4.6 Stochastic Analysis

A more accurate and complete understanding of the impact of dependencies on development can be obtained by means of a stochastic analysis with SDDA. In stochastic analysis, SDDA uses a probability density function of the punctuality, rather than a single instance. The corresponding output is a probability density function for the beginning and completion times of each system, accounting for the parameters of the model, and the overall effect of topology. The expected value of the beginning and completion times can be used as a first guideline to decisions, while the variance of these times gives insight into the risk associated with the development and delays. These outputs show patterns and features of the whole architecture. For example, given the expected distribution of punctuality, the user can compute the probability that development deadlines will be met. The user can then compare alternate architectures, identify their critical systems (i.e., systems whose delay cause the largest impact on the overall development), and identify the topological patterns that cause the observed criticality. Since SDDA is computationally not expensive (Table 7.9), Monte Carlo simulation appears to be the best choice to perform this type of analysis. The user generates a large number of instances of punctuality, based on given distributions, and then computes the expected beginning and completion times with SDDA.

The current implementation of SDDA uses the following model of uncertainty:

- The user inputs an expected level of punctuality for each system, as in SDDA deterministic analysis.
- The input level of punctuality will be the mode of a symmetric Beta Probability Density Function (PDF), multiplied by a spreading factor. This input

Table 7.9
Computational Cost of SDDA Analysis (Single Instance of Developmental Network. Processor Intel Core i3-2350M 2.3 Ghz)

Number of nodes	Average number of edges (100 runs)	Average time	Maximum time
100	2477	$2 \cdot 10^{-3}$ s	$2.7 \cdot 10^{-3}$ s
500	62374	0.0293 s	0.0297 s
1000	249703	0.108 s	0.109 s
2000	999548	0.412 s	0.432 s

level might not be the mean and median of the PDF, because the tails of the distribution might be cut, to respect the range of feasible punctuality.

- The user inputs one of three levels of uncertainty for each system (low, medium, and high). The lower the uncertainty, the more reliable the assumption of the punctuality. This higher confidence is modeled with a lower variance of the PDF.
- The user inputs the time instant, on the development schedule, at which SDDA has to compute the expected development performance. As this time instant gets closer to a system's completion time, the uncertainty on the system punctuality will decrease (lower variance of the PDF), until the chosen time is equal or greater than the system's completion time, at which point the uncertainty on the completion time is zero.
- If the Beta function resulting from the multiplication by the spreading factor, and the modified variance due to confidence and time instant is partially outside the allowed range of punctuality, it is cut over the range from 0 to 100, and normalized so that its area is equal to 1.

Figure 7.24 shows the results of development analysis via the stochastic SDDA model, over 10000 samples. The punctuality was generated according to the PDF of the stochastic SDDA model, and the consequent beginning and completion times of each system were computed based on the dependencies. The same stochastic model was also applied to PERT analysis. The matrices of strength and criticality of dependencies, minimum and maximum development times are the same as in the deterministic example of Figure 7.23. Punctuality of Nodes 1 and 3 is 100, punctuality of Node 2 is 70. Node 1 has medium uncertainty level, Node 2 has high uncertainty level, and Node 3 has low uncertainty level.

At time 0, the beginning and completion times show the largest uncertainty. At time 8 weeks, the uncertainty has decreased. Node 2 has higher uncertainty than Node 1, according to the input by the user. At time 16 weeks, two systems are fully developed, and therefore exhibit no uncertainty. Node 3 has low uncertainty, according to the input by the user. A development schedule according to the SDDA basic

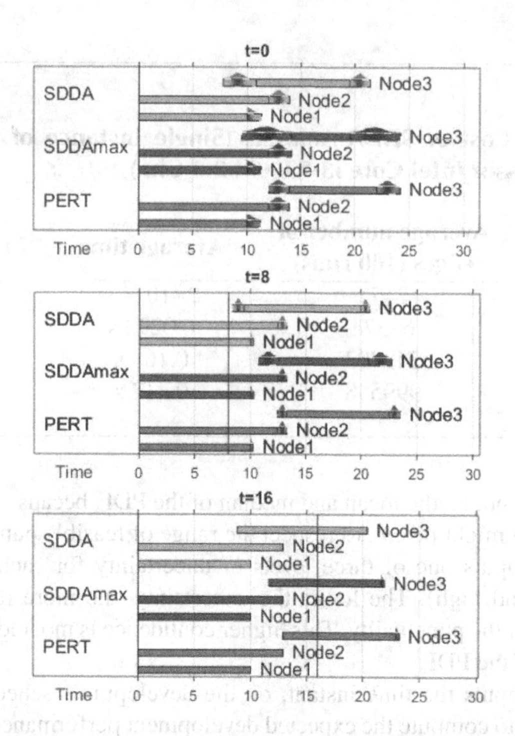

Figure 7.24: Gantt chart showing the schedule of development of the simple 3-node network shown in Figure 7.21, according to the SDDA, SDDAmax, and PERT stochastic models. The resulting PDF of beginning and completion times is shown above the corresponding bar of the Gantt chart. The darker shadows on the bars indicate the zones of higher probability. Top: analysis of expected schedule at time 0. Center: analysis at time 8 weeks. Bottom: analysis at time 16 weeks.

model shows that early completion is allowed, but the graph of the expected development at time 8 shows that there is still risk associated with the uncertainty. The user can make an informed decision thanks to this methodology, choosing an appropriate beginning time, based on result of this analysis, and on the amount of accepted risk. For example, the mean of the PDF of beginning time can be used as scheduled value. The user can also calculate the probability that each system will be fully developed by given deadlines. In this example, the expected levels of punctuality did not change. Of course, these levels, as well as the levels of uncertainty, may change over time. The user can repeat the analysis later, during the development of the complex system, when further decisions are required. SDDA results suggest possible times at which to perform the analysis again, with current information. Table 7.10 lists some of the results of stochastic analysis with SDDA, SDDAmax, and PERT model, including the expected completion time, and the 10th, 50th, and 90th percentiles of the beginning time, representing more or less conservative choices. The results are computed based on the information available at time 0. The values of the parameters and inputs are the same as in the previous cases.

Table 7.10
Results of Stochastic SDDA Analysis of a Simple 3-Node Network

Delayed system	Model	10th, 50th, and 90th percentile of t_B^3	$E\left(t_C^i\right)$ (weeks)
None	SDDA	[7.98 8.33 8.80]	[10.32 12.26 19.76]
	SDDAmax	[9.60 10.04 10.80]	[10.32 12.26 20.24]
	PERT	[12.04 12.23 12.54]	[10.32 12.26 22.37]
1 (P=70)	SDDA	[9.21 9.76 10.34]	[11.80 12.26 19.98]
	SDDAmax	[9.66 10.16 10.86]	[11.80 12.26 20.33]
	PERT	[12.06 12.27 12.58]	[11.80 12.27 22.40]
2 (P=70)	SDDA	[8.55 9.12 9.71]	[10.32 12.90 20.40]
	SDDAmax	[10.64 11.65 12.67]	[10.32 12.90 21.76]
	PERT	[12.48 12.91 13.33]	[10.32 12.91 23.01]
3 (P=70)	SDDA	[7.98 8.33 8.79]	[10.31 12.26 19.80]
	SDDAmax	[9.60 10.04 10.79]	[10.32 12.26 21.33]
	PERT	[12.04 12.22 12.53]	[10.32 12.26 23.46]
All (P=50)	SDDA	[11.69 12.36 13.02]	[13.00 13.50 24.35]
	SDDAmax	[12.19 13.12 13.93]	[12.99 13.50 25.09]
	PERT	[13.15 13.54 13.94]	[13.00 13.50 25.54]

As expected, PERT exhibits the highest final expected completion time $E\left(t_C^3\right)$, because no development overlapping is considered between dependent nodes. On the other hand, the percentiles show a smaller spread in the values of beginning and completion times in PERT (Table 7.10 shows percentiles of the beginning development time of Node 3, on 10000 instances). The larger spread in the values of beginning and completion times in SDDA and SDDAmax occurs because the uncertainty in these values is not only due to the stochastic nature of the punctuality (as it happens in PERT), but also to the *lead time* assigned to the systems, based on the parameters of the models and on the expected punctuality.

7.1.4.7 Source of Parameters

This section provides the user with guidelines about the different possible sources of parameters to model systems behavior with SDDA.

Historical Data This kind of approach is based on historical data, or knowledge database. Keeping into account the meaning of the parameters, the user evaluates them based on existing information about partial parallel development of system, and punctuality confidence levels.

Input Requirement-Based Approach An alternative data-based approach requires the user to estimate the amount of input that a receiver node requires from its feeder nodes before its own development can start. This input will suggest an appropriate SOD. Then, based on the specific problem and on the amount of risk that can be accepted, the user will evaluate the COD. Guidelines to estimate the amount of required input can be found in literature, for example [114].

Expert Judgment-Based Approach Based on the intuitive meaning of the two parameters of SDDA model for a one-to-one dependency, an expert user can directly input appropriate parameters into the model. Figure 7.22 is useful to understand the relationship between completion time of a feeder node and beginning time of a receiver node in function of punctuality, when the parameters of the model change. In the case of expert judgment-based approach, I strongly recommend to perform an evaluation of the sensitivity of the results of SDDA analysis to changes in the parameters.

7.1.5 EXAMPLE OF APPLICATION OF SDDA: A COMMUNICATION SATELLITE

This section illustrates a simple use of SDDA to quantify high-level systems criticalities and delay absorption in the development of a communication satellite. For this analysis, the redundant nodes have been condensed in individual categorical nodes. The developmental dependencies of the systems on board of the satellite do not generally coincide with the operational dependencies. Figure 7.25 shows the component systems and their developmental dependencies for this application.

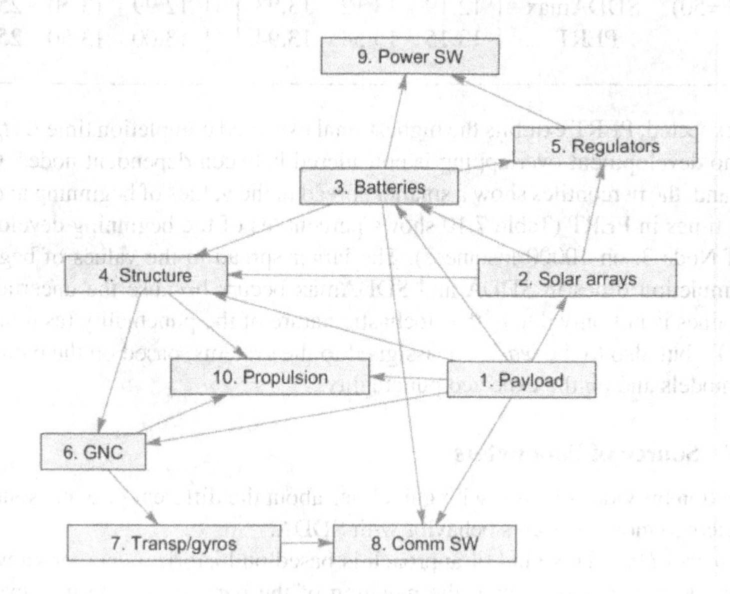

Figure 7.25: Developmental dependencies of systems on board a communication satellite.

As described in Subsection 7.1.4, developmental dependencies in SDDA are similar to dependencies in a PERT network, but SDDA models not absolute dependencies, allowing for partial overlapping in the development schedule. Since a system can begin its development with a *lead time*, the whole development process can absorb part of the delays that might occur in the development of some of the systems involved. The values of strength and criticality of the developmental dependencies in the communication satellite are as follows (the node numbers represent the modules in Figure 7.25):

$$
SOD = \begin{array}{c} \\ \mathbf{1} \\ \mathbf{2} \\ \mathbf{3} \\ \mathbf{4} \\ \mathbf{5} \\ \mathbf{6} \\ \mathbf{7} \\ \mathbf{8} \\ \mathbf{9} \\ \mathbf{10} \end{array}
\begin{array}{c} \begin{array}{cccccccccc} \mathbf{1} & \mathbf{2} & \mathbf{3} & \mathbf{4} & \mathbf{5} & \mathbf{6} & \mathbf{7} & \mathbf{8} & \mathbf{9} & \mathbf{10} \end{array} \\
\begin{bmatrix}
0 & 0.4 & 0.5 & 0.25 & 0 & 0.15 & 0 & 0.7 & 0 & 0.3 \\
0 & 0 & 0.5 & 0.3 & 0.7 & 0 & 0 & 0 & 0 & 0 \\
0 & 0 & 0 & 0.2 & 0.6 & 0 & 0 & 0.5 & 0.7 & 0 \\
0 & 0 & 0 & 0 & 0 & 0.4 & 0 & 0 & 0 & 0.5 \\
0 & 0 & 0 & 0 & 0 & 0 & 0 & 0 & 0.3 & 0 \\
0 & 0 & 0 & 0 & 0 & 0 & 0.7 & 0 & 0 & 0.4 \\
0 & 0 & 0 & 0 & 0 & 0 & 0 & 0.45 & 0 & 0 \\
0 & 0 & 0 & 0 & 0 & 0 & 0 & 0 & 0 & 0 \\
0 & 0 & 0 & 0 & 0 & 0 & 0 & 0 & 0 & 0 \\
0 & 0 & 0 & 0 & 0 & 0 & 0 & 0 & 0 & 0
\end{bmatrix} \end{array}
$$

$$
COD = \begin{array}{c} \\ \mathbf{1} \\ \mathbf{2} \\ \mathbf{3} \\ \mathbf{4} \\ \mathbf{5} \\ \mathbf{6} \\ \mathbf{7} \\ \mathbf{8} \\ \mathbf{9} \\ \mathbf{10} \end{array}
\begin{array}{c} \begin{array}{cccccccccc} \mathbf{1} & \mathbf{2} & \mathbf{3} & \mathbf{4} & \mathbf{5} & \mathbf{6} & \mathbf{7} & \mathbf{8} & \mathbf{9} & \mathbf{10} \end{array} \\
\begin{bmatrix}
0 & 20 & 40 & 15 & 0 & 30 & 0 & 25 & 0 & 30 \\
0 & 0 & 50 & 30 & 10 & 0 & 0 & 0 & 0 & 0 \\
0 & 0 & 0 & 10 & 35 & 0 & 0 & 20 & 15 & 0 \\
0 & 0 & 0 & 0 & 0 & 40 & 0 & 0 & 0 & 30 \\
0 & 0 & 0 & 0 & 0 & 0 & 0 & 0 & 15 & 0 \\
0 & 0 & 0 & 0 & 0 & 0 & 10 & 0 & 0 & 25 \\
0 & 0 & 0 & 0 & 0 & 0 & 0 & 5 & 0 & 0 \\
0 & 0 & 0 & 0 & 0 & 0 & 0 & 0 & 0 & 0 \\
0 & 0 & 0 & 0 & 0 & 0 & 0 & 0 & 0 & 0 \\
0 & 0 & 0 & 0 & 0 & 0 & 0 & 0 & 0 & 0
\end{bmatrix} \end{array}
$$

The minimum and maximum development times of the systems are (in weeks)

$$
t_{min} = \begin{bmatrix} 14 \\ 12.5 \\ 6 \\ 11 \\ 3 \\ 2.5 \\ 15 \\ 4 \\ 3 \\ 9 \end{bmatrix} \qquad
t_{max} = \begin{bmatrix} 24 \\ 17 \\ 12 \\ 23 \\ 5 \\ 4 \\ 21 \\ 5 \\ 4 \\ 12 \end{bmatrix}
$$

Figure 7.26 shows the effect of partial developmental dependency. While PERT considers all the dependencies to be absolute, meaning that each node must wait until all its predecessors are fully developed, SDDA and SDDAmax consider partial overlapping based on the parameters that model the dependencies.

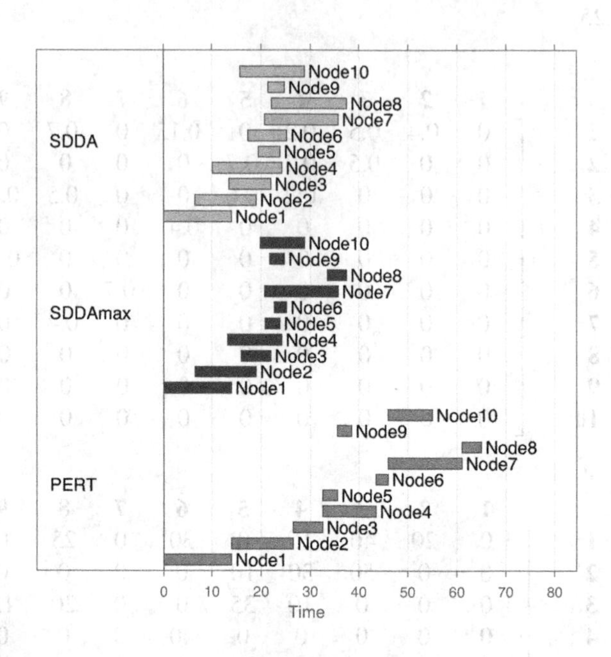

Figure 7.26: Gantt chart of the development of a communication satellite, comparing SDDA model, SDDAmax model, and PERT.

Results show how the development can be completed in 37.5 weeks, when no delays arise, while PERT computation resulted in 65 weeks expected for development time, due to the absolute dependency between systems. However, the figure also shows how early development causes a few systems, including the GNC (node 6), the propulsion system (node 10) and the communication software (node 8), to have to wait for the completion of other systems, extending their own development over a longer time span. While this can result in early achievement of partial capabilities and increase the possible absorption of delays, it also causes an increase in cost. Decision-makers must trade-off between these competing aspects.

7.1.5.1 Delay Absorption

Besides the baseline schedule, SDDA analysis can also be used to quantify the delay absorption throughout the development of a complex system, and to identify the most critical elements in terms of impact on the development time. When the development of a system slows down, the delay affects the development of other systems in the network. In PERT, the delay in a system reverberates entirely on the development of the dependent systems. Delays on the critical path cannot be recovered. If the dependencies are not absolute, i.e., only part of the development of a system depends on the predecessor, the development process of this system can be partially executed even when it is lacking some of the inputs from the systems on which it depends. This results in both a faster overall development, and in possible partial delay absorption. Table 7.11 lists the results of deterministic SDDA analysis, when one system at a time experiences a delay, with its punctuality down to 30, while the other systems are all keeping their fastest development.

Table 7.11
Delay Absorption in the Development of the Communication Satellite

Delayed node	SDDA		SDDAmax		PERT	
	Final time (weeks)	Delay recovery (%)	Final time (weeks)	Delay recovery (%)	Final time (weeks)	Delay recovery (%)
Payload	51.1	0	51.1	0	72.0	0
Solar arrays	40.7	0	46.5	0	68.2	0
Batteries	39.6	50	48.5	0	69.2	0
Structure	42.8	37.1	47.4	0	73.4	0
Regulators	37.5	100	37.5	100	65	100
GNC	41.4	0	42.1	0	66.1	0
Transp/gyros	41.7	0	43.3	0	69.2	0
Comm SW	37.8	55	38.2	0	65.7	0
Power SW	37.5	100	37.5	100	65	100
Propulsion	37.5	100	37.5	100	65	100

Results show how the partial overlapping of development schedule in SDDA allows for total or partial delay recovery in nodes that are along the critical path (when PERT does not recover any delay). The more conservative SDDAmax, which has less partial overlapping of development schedule, exhibits less delay recovery. However, this recovery is measured as a percentage with respect to the delay in the affected node. Even if this particular delay causes less overlapping and it is not absorbed, especially in the case of critical dependencies, the final development time with the SDDAmax model is still shorter than the final development time that the complex system would have due to absolute developmental dependencies. SDDA model indicates that delays in the development of the payload heavily impact the overall development. Delays in the development of the structure, GNC system and transponder/gyros cause some delay in the final development. Thanks to partial parallel development, however, delays in the development of the propulsion system and of the software can be partially or completely absorbed.

Figure 7.27 shows one application of stochastic analysis, according to the methodology described in Subsection 7.1.4. The development uncertainty of batteries, structure, communication software and power system software is low. The development uncertainty of payload, regulators, transponder/gyros and propulsion is medium. The development uncertainty of solar arrays and GNC is high. The uncertainty in the completion time of the systems causes uncertainty on the "best" beginning time of other systems. The user can use this information to decide when to begin the development of each system, based on current information and amount of accepted risk. The expected value of completion under this scenario is 39.9 weeks in SDDA, 41.3 weeks in SDDAmax, and 66.9 weeks in PERT.

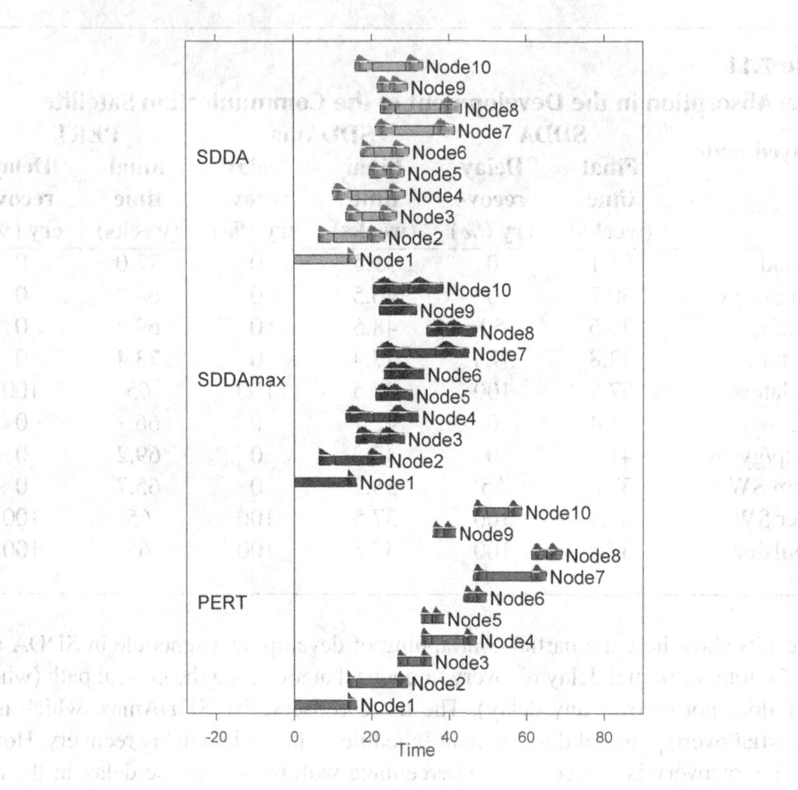

Figure 7.27: Gantt chart for communication satellite under uncertainty. Information at time equals 0.

Figure 7.28 shows the reduction in uncertainty when the analysis is conducted again after 20 weeks. The improved information decreases the uncertainty, and the resulting expected value of completion under this scenario is 38.2 weeks in SDDA, 38.8 weeks in SDDAmax, and 66.2 weeks in PERT.

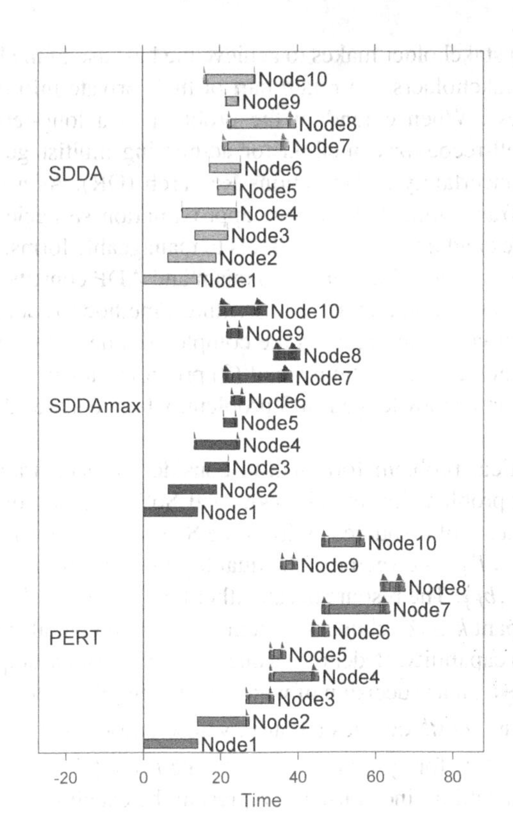

Figure 7.28: Gantt chart for communication satellite under uncertainty. Information at time equals 20.

7.1.6 MULTI-STAKEHOLDER DYNAMIC OPTIMIZATION

SoSs are characterized by operational and managerial independence. Stakeholders involved in a design pursue individual objectives that often conflict with each other and with the objective of the entire SoS.

Many conflicts stem from the competition for limited resources. A top manager with partial authority is more common than a centralized authority; therefore, mechanisms are required that can influence the behavior of the participating stakeholders. It is also crucial to include both the near-term and long-term objectives in the early design phase, to improve the capability or reduce the cost in the long run. The Multi-Stakeholder Dynamic Optimization (MUSTDO) [61] methodology supports coordination of conflicts via an integration of transfer contract mechanism and Approximate Dynamic Programming (ADP).

A transfer contract is defined as the compensation that each participating stakeholder needs to pay to other stakeholders for consuming the shared resources provided by the top-level manager. The transfer contract can influence the decisions that each participating stakeholder makes to achieve the best use of the limited resources. Meanwhile, the stakeholders can retain part of their private information during the negotiation process. When extending the problem to a long-term horizon, ADP [20, 141] is a well-recognized method for addressing multistage decision-making problems under uncertainty in Operations Research (OR). As a flexible modeling and algorithmic framework with various approximation strategies, ADP excels in converting complex and intractable problems to manageable forms. The combination of the transfer contract coordination mechanism and ADP constitutes the MUSTDO method. This method is valuable in: (1) providing a method to address the multistage composition decisions, which reduces the complexity due to the increasing number of systems and amount of uncertainty, and (2) providing a new perspective to coordinate conflict in an acknowledged SoS problem, which enables the efficient use of resources.

The mathematical problem formulation considers a decentralized dynamic resource allocation problem for an acknowledged SoS. A group of SoS participants $\mathscr{K}$ competes for a set of resources $\mathscr{R}$ from the SoS manager over a finite time horizon $\mathscr{T} = \{1, 2, \cdots, T\}$. The vector of the quantity of each resource $r \in \mathscr{R}$ is denoted as $b = \{b_1, b_2, \cdots, b_r\}$. The resources can either be monetary value or physical entities. Each participant $k \in \mathscr{K}$ chooses systems from its potential system pool $\mathscr{I}^k$ to maximize its own capability. x_t^k denotes a decision vector of participant k at time step t. Each entry $x_{i,t}^k$ is a binary decision of whether a potential system $i \in \mathscr{I}^k$ is selected into the architecture. $cost_t^k$ denotes a matrix whose entries $cost_{i,r,t}^k$ represent the unit cost of resource $r \in \mathscr{R}$ for system $i \in \mathscr{I}^k$ at time t. cap_t^k is a vector whose entries $cap_{i,t}^k$ indicate a capability index used to represent the capability of each system.

7.1.6.1 SoS Manager's Problem

The goal of a *super* SoS manager (if one exists) with ultimate control over all the participants and systems is to maximize the entire SoS capability with the best use of available resources. In such an ideal case, the mathematical formulation is given as follows:

$$\underset{x_i}{\text{maximize}} \quad E\left[\sum_{t=1}^{T} cap_t \cdot x_t\right] \tag{7.47}$$

$$\text{subject to} \quad cost_t \cdot x_t \leq b_t \qquad \forall t \in \mathscr{T} \tag{7.48}$$

$$\sum_{i \in \mathscr{I}^k} x_{i,t}^k \leq 1 \qquad \forall t \in \mathscr{T}; k \in \mathscr{K} \tag{7.49}$$

$$x_{i,t} \in \{0, 1\} \qquad \forall i \in \mathscr{I}; \forall t \in \mathscr{T} \tag{7.50}$$

$$\text{transition function} \quad b_{t+1} = b_t - cost_t \cdot x_t \tag{7.51}$$

$$cap_{t+1} = cap_t + \widehat{cap}_{t+1} \tag{7.52}$$

Equation 7.47 is the objective function that seeks to maximize the expected SoS capability over the entire horizon. The SoS capabilities are achieved through a linear combination of capabilities from the selected systems, which is a simplification of the typical nonlinear capability functions in the real world.

Equation 7.48 describes the resource constraint that bounds the objective function at each time step. The resources are given at the first stage and are depleted by the participants' decisions as time proceeds. This evolution process is expressed in Equation 7.51. Equation 7.49 is the constraint that allows only one system under participant k to be acquired at each time step. This constraint is only to simplify the problem. Equation 7.50 is the binary constraint for the decision variables, where $x_{i,t} = 1$ means acquiring system i, while $x_{i,t} = 0$ means not acquiring system i. Equation 7.52 describes the system capability improvement process over time, where $\widehat{cap}_{t+1}$ is the change of capability. The capability change can result from the technology maturation, policy change, and so on.

The solution in this ideal centralized case serves as a benchmark of the values that the SoS manager in the decentralized case hopes to achieve or at least approach. In other words, the problem of the SoS manager is to design a mechanism (for example, a transfer contract coordination mechanism) that can guide the participants toward the SoS-level optimal capability.

7.1.6.2 SoS Participant's Problem

The objective of each participant in an SoS is to determine a sequence of decisions that can maximize the expected sum of individual capability over time, given the mechanism and resources provided from the SoS manager. The mathematical program is given by the following equations:

$$\underset{x_i^k}{\text{maximize}} \quad E\left[\sum_{t=1}^{T}\left(cap_t^k \cdot x_t^k + TC_t^k\right)\right] \tag{7.53}$$

$$\text{subject to} \quad cost_t^k \cdot x_t^k \leq b_t - \sum_{j \neq k} cost_t^j \cdot est^{\cdot} x_t^j \qquad \forall t \in \mathcal{T} \tag{7.54}$$

$$\sum_{i \in \mathcal{J}^k} x_{i,t}^k \leq 1 \qquad \forall t \in \mathcal{T}; k \in \mathcal{K} \tag{7.55}$$

$$x_{i,t}^k \in \{0,1\} \qquad \forall i \in \mathcal{J}; \forall t \in \mathcal{T} \tag{7.56}$$

$$\text{transition function} \quad b_{t+1} = b_t - cost_t x_t^k - \sum_{j \neq k} cost_t^j \cdot est^{\cdot} x_t^j \tag{7.57}$$

$$cap_{t+1}^k = cap_t^k + \widehat{cap}_{t+1}^k \tag{7.58}$$

Equation 7.53 describes the objective of participant k that maximizes the expected sum of system capabilities with the incorporated transfer contract coordination mechanism over the entire horizon. System i provides to participant k's portfolio the same capability contribution as it does to the ideal centralized case for the SoS manager. TC_t^k is a shorthand notation of the transfer contract coordination mechanism that follows the SoS manager's guide. Equation 7.54 delineates the resource constraint that bounds the objective function at each time step. This constraint has two features: first, the resources are provided at the initial stage and are depleted over time by executed decisions, as described in Equation 7.57. Resources can be replenished over time if needed. Second, all the participants are constrained by the shared resources. Since one participant does not know the decisions of other participants, $est^{\cdot}x_t^j$ is used to represent participant k's estimation of participant j's decisions on the selection of systems. Therefore, the available resources for participant k are the total resources minus its own past consumption and the estimation of other participants' resource consumption. Equation 7.55 is the constraint that only allows one system under participant k to be acquired at each time step. Equation 7.56 is the binary constraint. Equation 7.58 gives the evolution of capability cap_t^k by adding the uncertain capability change $\widetilde{cap}_{t+1}^k$.

The problem of each participant is to solve the individual optimization and communicate with each other on the transfer contract. The decisions made by the independent participants should be able to produce total capability that is close to the optimal capability of the SoS manager in the ideal centralized case.

7.1.6.3 Transfer Contract Coordination Mechanism and Approximate Dynamic Programming

The transfer contract, or transfer pricing coordination mechanism, was originally developed as a tool for revenue management in multi divisional corporations, where the separate divisions are autonomous profit centers [96]. The transfer contract coordination mechanism deals with the problem of pricing the products and services that are exchanged between such divisions within a firm. As illustrated in Figure 7.29, a transfer price represents the price for the internal market when business unit B needs a product from business unit A. An appropriate transfer price for the internal exchange is required to achieve the firm-level optimal profit. The essential idea of the mechanism is to align the objective of individual divisions to that of the firm by encouraging efficient sharing of resources.

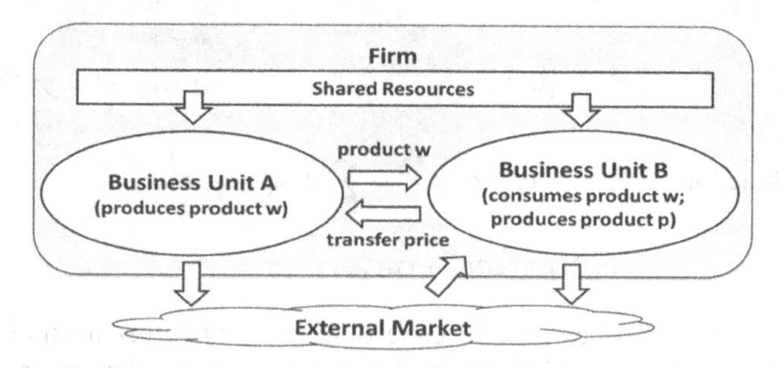

Figure 7.29: Transfer pricing in decentralized corporation.

The interpretation of a transfer contract varies based on the context but, in essence, the effect of a transfer contract is to change the objectives and behaviors of individual participants.

When extending this approach to multistage decision-making in a long-time horizon, the principle still applies, with the caveat that the influence of future decisions and values will complicate the calculation process. The use of ADP simplifies the problem for the stakeholders. ADP utilizes an approximation of the value function for making decisions, which alleviates the burden of collecting complete information to build a perfect model for optimizing a complex SoS.

Figure 7.30 summarizes the workflow of the MUSTDO method for decision-makers. The vertical line represents the time evolution. To start, the SoS manager announces the constrained resources and the MUSTDO method to the SoS participants. The SoS manager only has partial authority to enforce the collaboration. SoS participants k and j negotiate with each other based on the rules in the MUSTDO method. However, the participants are unlikely to reveal all information to each other for political, economic, or technology-related reasons, which explains the independence of each participant's decision-making (a trait of SoS). In the central part of Figure 7.30, the circles represent the architectural states of the system composition for individual participants, while the diamonds represent decision nodes where the negotiation and decisions happen. The decision-making process happens on the decision node and is displayed in the rounded rectangular call-out. The process produces, for each participant, a final sequence of decisions, though each participant only executes the decisions for the current time stage. After the participants execute the current decisions, new information such as system capability and cost change from the external environment may update the approximate value functions, which affects the decisions in the next stage.

The procedure in the call-out can be considered as a simulation model that forecasts the sequential decision-making that will take place. Participants k and j accept to use the MUSTDO method. First, each participant proposes a set of potential systems to acquire, giving partial information to the SoS manager and to other participants. Each participant determines the structure and basis functions for its future value function approximation (e.g., linear function, piece-wise linear function), based on the goals and past observations of the problem features. For example, a quasi-linear relationship between the total amount of resources and the aggregated SoS capabilities in the past could indicate a linear approximation curves for the practitioners. In addition, each participant evaluates the uncertain information (e.g., fluctuation of system capability) in the form of probability distributions or simulation models that might affect the parameters. After the preparation step, each participant initiates the coefficients of the basis functions and exchanges the initial transfer contract. Then, the participants run their own dynamic optimization problems. During the process, the value function and transfer contract are updated based on the samples from the given distribution or simulation model. These transfer contracts need to be communicated to other participants. The communication is an automated process; however, the decision-makers are free to be involved in the communication.

The process continues until the approximate value of the future capability converges. The final solution contains a sequence of acquisition decisions over sequential stages, associated capability, and transfer contracts based on different realizations of uncertainty. During the whole implementation process, the SoS manager does not need to be involved in the negotiation procedure.

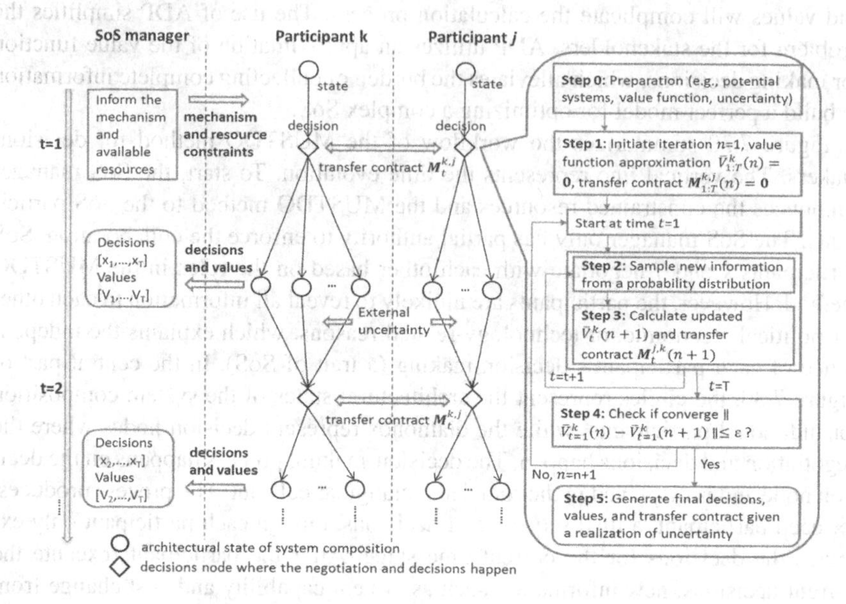

Figure 7.30: Workflow of the MUSTDO framework.

7.2 OTHER USEFUL METHODS FOR SOS MODELING AND ANALYSIS

Researchers developed many other tools and methods or modified existing tools and methods towards SoS Modeling and Analysis. This section provides a brief overview of some of these methods.

7.2.1 SYSTEM DYNAMICS

System Dynamics (SD) is a modeling method which provides the ability to articulate non-linear cause-effect behavior among system components by studying the positive and negative relationships between them and the feedback loops which lead to emergent behaviors. SD is comprised of individual systems, feedback loops, stocks, and flows. SD represents the dynamics of complex SoS by using the Casual Loop Diagram (CLD) and the Stock and Flow Diagram (SFD). Together, these diagrams provide the ability to qualitatively and quantitatively analyze the dynamics of complex SoS including traits such as interdependence, circular causality, mutual interaction and information feedback.

The CLD gives a general representation of the structure of the system and insight into the interdependencies within the system and how these interdependencies change with time to contribute to the system as a whole. The main components of the CLD are known as feedback loops and come in two types – positive, or reinforcing, and negative, or balancing.

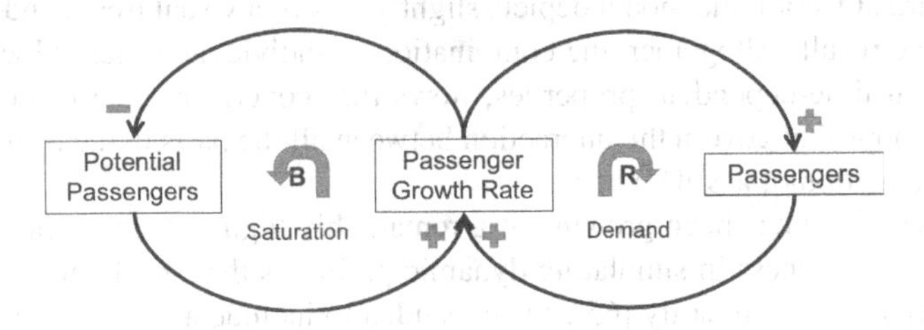

Figure 7.31: Causal loop diagram for air transportation passengers.

Figure 7.31 demonstrates a simple demand modeling example in which two governing feedback loops determine the relation between their encompassed properties of market saturation and demand, and these two properties solely govern the state of passengers. The "demand loop" or "R" loop is the positive correlation between demand and the number of passengers in the network; here the "+" signs show the positive correlation between the rate of passenger demand growth and the total number of passengers who travel. The "saturation loop" or "B" loop is the negative correlation and balances the total passengers who travel in the network. Thus, as the passenger growth rate increases, the potential number of passengers decreases; thus the "-" shows the negative correlation.

The SFD provides the ability for more detailed quantitative analysis. SFD applies a quantitative aspect to the CLD in the form of blocks known as stocks. Stocks are system-dependent dynamic entities, which can be either individual systems or a property of interest in the whole system. Stocks are defined by their dependence on time, and as so they accumulate or deplete with time. The rate of that change is defined as the flow.

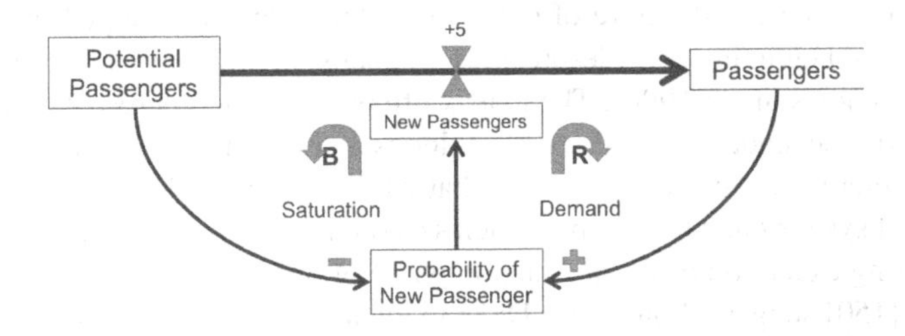

Figure 7.32: Stock and flow diagram.

In the very simple SFD example shown in Figure 7.32, the CLD is adapted to display more quantitative results. The Potential Passengers and Passengers are now replaced by stocks, which have dynamic numerical values that change with each time step. They are connected by a single flow in the direction of the Passenger stock, which indicates that 5 new passengers are entering the Passenger stock at each time step, and 5 Potential Passengers are leaving. The relationships from the feedback loops still exist, but some of the encompassing loops are replaced by their stock/flow counterparts. Overall, the model depicts slightly less of a visual trend, and more of a quantitative result. Altogether, the combination of individual system blocks, stocks that contain time-dependent properties, flows that govern the rate of change, and feedback loops that govern the interaction between all the parts in the overall system are the foundation of a SFD model.

The use of SD has been growing at a remarkably high rate as people realize its ability and uniqueness in simulating dynamic problems that are dominated by their complexity and non-linearity [65]. SD expanded to include analysis in engineering, politics, land use planning, US Space Program, and more. SD is best used when system boundaries are clearly identified; this would help in effective causal loop diagram [64]. Once the constituent systems of an SoS are identified, these can be represented as stocks, and the interactions between them as the links, to identify how these systems affect one another. SD can also help policy-makers to better understand impacts of their strategies and policies on SoS performance. SD's utility to SoS analysis comes from the fact that it makes it very easy to observe rippling effects by analyzing influences from feedback loops. However, one of the downfalls of this method is that without sufficient experience, it is easy to fall prey to incorrect insights to problems [57]. A typical downfall is that incorrect stocks, and particularly feedback loops and flows are assigned to portions of the diagram, resulting in an inaccurate flow of information throughout the system, which is the foundation of a proper SD.

7.2.2 DESIGN STRUCTURE MATRIX

The Design Structure Matrix (DSM) is a simple, compact and visual representation of connections between entities; for example, component systems in an SoS in the form of a square matrix. DSM is used in systems engineering and project management to model the structure of complex systems [56]. Developed in the 1960s, DSM saw the beginning of widespread use in many engineering industries and government agencies in the 1990s. The matrix offers a compact representation of information about dependencies and feedback loops and modules. In SoS Modeling and Analysis, there are various algorithms that can be used to identify the underlying structure of systems and their dependencies (modules, paths, and loops). This is very useful during execution of the abstraction phase of SoS Modeling and Analysis. For example, [150] suggests how to use DSM to identify complexity in SoS.

7.3 CHAPTER SUMMARY

SoS have features that cannot always be treated with tools from other disciplines. In this chapter we presented some tools that specifically address SoS features.

The key ideas presented in this chapter were:

1. SoS features are unique and, at the same time, very diverse. A single tool to address all of these features would be impractical.
2. The Analytic Workbench is a suite of tools which use a common representation of SoS as a network of systems and dependencies. Each tool addresses a different aspect of SoS.
3. Existing tools can also be modified to address SoS features.
4. The user has to clearly define goals and objectives, in order to decide the appropriate tool(s) to use for SoS Modeling and Analysis.
5. SoS-specific tools address various hierarchical levels, and provide support to decision-making in conditions of uncertainty.
6. Optimization in SoS still faces many challenges, but SoS tools provide information, especially to downselect architectures in early stage of systems design.

8 Enhancing System of Systems Engineering

8.1 ARTIFICIAL INTELLIGENCE, MACHINE LEARNING, AND AUTONOMY

The "Solberg Chart" (Figure 1.1) is a snapshot in time of the concepts and methodologies useful for the study of SoS. This chart shows several waves associated with Artificial Intelligence (AI), whose products also appear as Automata, Neural Networks, Swarm Intelligence, etc. The term Machine Learning (ML) was not yet in use at the time but could easily be inserted in many of these streams. Nowadays, as AI/ML advances into more and more applications, including through semi- or fully-autonomous systems as components in SoS (Chapter 6), a clear understanding of how AI/ML enhance our ability to perform SoS Modeling and Analysis and drive important architecture design decisions is essential.

But before examining "intelligence," the more basic notion of autonomy is quite obviously important in SoS. First, the essence of what characterizes SoS is the *independence* of the components, which means that each participant in an SoS has some degree of autonomy. Additionally, one of the three classes of SoS design variables used in the abstraction phase (Section 3.4) is *control*, referring to how much control, therefore autonomy, an SoS participant has and how the amount of control changes (or should change) depending on circumstances. Machines all around us, including everything from military aircraft to airport shuttles to vacuum cleaning robots, are becoming semi- or fully autonomous. Therefore, the now very wide field of autonomy, which drives traditional areas like control and robotics, has a connection with SoS modeling and analysis, and we need to thoroughly and explicitly include considerations on autonomy in the architecture design of SoS.

In Part 8 of the SEBoK [24], there is a very nice introductory article on the intersection of AI/ML and SE, including identification of two streams: Artificial Intelligence for Systems Engineering (AI4SE) and Systems Engineering for Artificial Intelligence (SE4AI). In AI4SE, or in our case AI4(SoS)E, performing the complex modeling, analysis and decision-making required in SE, SoSE, and even Mission Engineering (ME) can be enhanced by AI/ML and the impressive abilities to consider large decision/search spaces, navigate uncertainty, and make constantly improved decisions through learning. SE4AI, on the other hand, seeks to bring key activities historically in the bailiwick of SE, such as requirements definition and attainment tracking, verification/validation, and a host of others, to the design and realization of AI/ML systems especially in safety or security intensive applications.

Among the very many observed and potential intersections of AI/ML and SoS, we will briefly explore a couple that fall in the AI4(SoS)E category and in fact are often related to each other:

- *AI/ML to drive SoS (or SoS component's) behavior*. This aspect is related to the independence of participating systems and their autonomous decision-making and degree of control. Methodologies from Agent-Based Modeling (Chapter 6) and BKI agent template (Figure 6.2) are completely compatible with incorporation of AI/ML and learning to drive complex, adaptive behavior. The SoS architect can identify goals of autonomy in SoS components' model, test, and design appropriate levels of control, and assess potential emergent behavior due to AI within the SoS.
- *AI/ML to extract and analyze data*. This aspect is related to the use of AI/ML methods and tools to analyze the outcome of SoS simulations or actual operations, to identify patterns, and to associate them to specific SoS features that can be appropriately leveraged. This use of AI/ML, which often provides valuable information to the design of intelligent systems, in particular utilizes the already vast family of machine learning algorithms and forms of "intelligent query agents."

8.1.1 AI/ML AS DRIVER AND ANALYZER

Even just scanning the SoS application domains shown in Table 4.2, it is not hard to identify ways in which AI/ML are already, today, driving behavior of key human/organizational and machine SoS components. This is driven in many cases by increasing complexity of the decision search space SoS components must navigate. AI can provide valuable contribution to the design and operation of SoS by enhancing behavioral decision-making and analysis of outcomes for planning of evolved SoS. AI/ML provides powerful means to understand and identify patterns and at the same time infuse AI inside the SoS itself, to evolve it in a smart way. A research project completed by Purdue University's CISA group under the DARPA Gamebreaker program (documented in [39] and [49]) provides a good example of just this setting. Before describing why this is the case, it is noteworthy that this DARPA program was a joint effort between the Agency's Artificial Intelligence Exploration (AIE) initiative and the Strategic Technologies Office (STO). STO is where most of the DARPA SoS research has taken place.

The project took SoS M&A concepts to the Mission Engineering (ME) level and created the Learning to Gamebreak (L2G) framework, a continuous-learning *SoS engine* that can manage the SoS and evolve it towards its goal. Perhaps the easiest way to understand this newly emerging concept of ME is to reason as follows: just as SoS is a form of meta-design (above the design of monolithic systems), ME is meta-design above the singular SoS level. Essentially, what portfolio of SoS capabilities is required to satisfy (robustly) a set of uncertain mission threads under a general family of missions. Design in a ME context represents a challenging design problem involving very large design spaces compared to other systems design problems, with

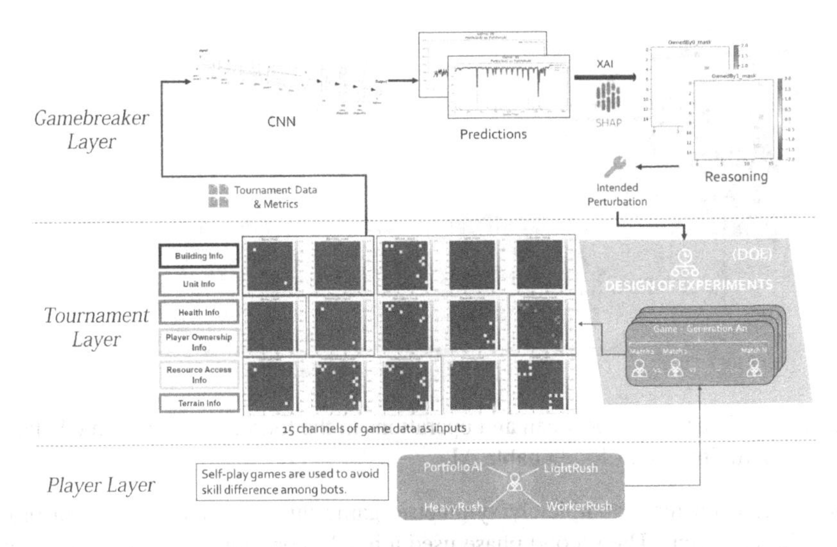

Figure 8.1: L2G framework for use of AI/ML in automated game balancing.

dynamic interactions between the mission environment and multiple system(s)-of-systems and their self-interested agent(s) operating in that environment, and with continuous evolution of environmental features, agent characteristics, or both.

The DARPA program challenged performers to explore the concept of "game balance" and whether such balance could be measured accurately in Real-Time Strategy (RTS) games and, further, turned into optimal means for unbalancing the game. In the very large and lucrative world of commercial RTS games, game developers want to achieve game balance – where balance implies that players at a variety of skill levels can still enjoy an extended and pleasing experience playing the game. As perhaps we all know from our childhood, playing a game in which you lose (or even win!) immediately is not very fun. The L2G framework carefully applies appropriate AI/ML techniques to existing RTS video games to quantitatively assess game balance, identify parameters that significantly contribute to the game balance, and explore new capabilities, tactics, and rule modifications that are most unbalancing to the game. Game balance in the context of this project is measured through the expected win/loss ratio of players of equal skill level, for which an appropriate model has been built. As mentioned, the commercial gaming industry has a long-standing interest in maintaining game balance since balanced games are more entertaining (and thus more popular), and market pressure drives their development. Contrary to these market goals, strategists in the military deliberately explore technologies that maximize imbalance to increase the probability of their winning. This project also contributed to close this existing gap in AI and data analytics research as applied to current war-gaming and simulation.

The first phase of the project utilized a simple gaming research environment to uncover interesting observations, for example the emergence of an initial game

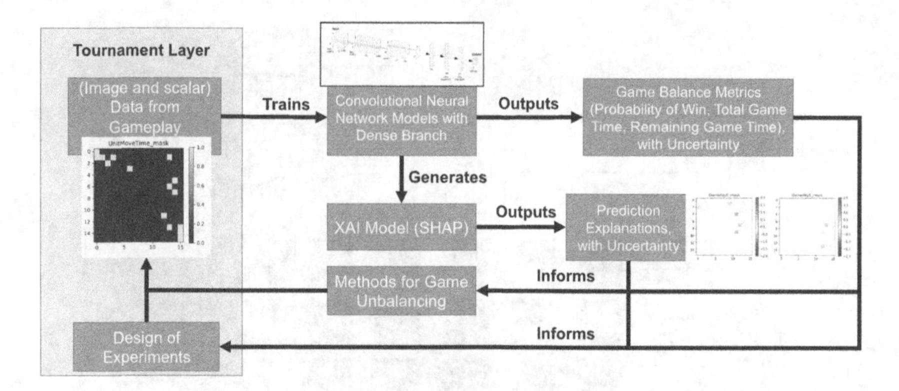

Figure 8.2: Interactions between and operations in the tournament and game balance layers, including use of Explainable AI.

imbalance even for a completely symmetric game due to certain movement order logic for the agent. The second phase used a much more complex gaming environment via use of the StarCraft II game. The whole framework (shown in Figure 8.1) brought together a unique combination of Design of Experiments (DoE) [142] for specifying simulations for the chosen agents, Machine Learning for predicting the outcomes of the game with desirable accuracy, Uncertainty Quantification (UQ) [143] for improving the robustness of the models, and Explainable AI (XAI) [89] to identify key features that impact game balance. Optimal learning has been employed for efficiently iterating upon the DoE to obtain better predictions and explanations for the game balance.

Mapping back to our AI4(SoS)E context, in this case explainable AI is used *outside* the SoS (in this case residing in the RTS Game Tournaments), to identify patterns and influence future behavior of the SoS, while SoS component behavior-generating AI is used *inside* the SoS via AI game bots, to drive autonomous agents acting naturally to pursue their goals. This interplay is illustrated in Figure 8.2.

8.1.2 AI/ML FOR EXTRACTION AND ANALYSIS OF DATA

SoS are defined by their architecture and by the interaction between the component systems. As described in Chapter 2, emergent behavior produced by these interactions is a fundamental characteristic of SoS, and it is important to identify and model the emergence as early and accurately as possible. To help the architecture design of SoS, it is also useful to discover patterns of "good" or "bad" behavior, that is architectural features associated respectively with high or low performance, according to appropriate user-defined metrics. This extraction of data can be used in two different ways:

1. Evaluation of SoS patterns and features.
2. Gathering of information to provide input to SoS Modeling and Analysis tools.

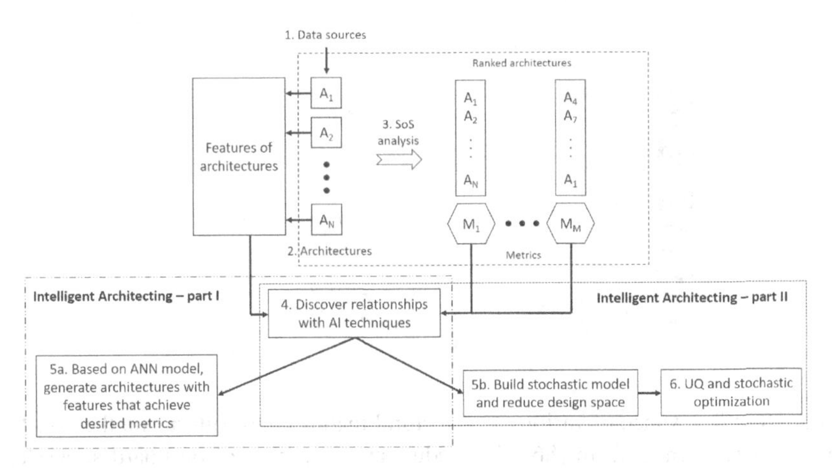

Figure 8.3: Use of AI products to evaluate positive patterns and features of SoS architectures.

In the first type of application, existing or planned architectures are evaluated with data evaluation AI tools, such as Neural Network, to extract information about architectural patterns that are associated with desired behavior. For example, [85] and [86] apply Bayesian-regularized deterministic Neural Network and fully stochastic Bayesian Neural Network to identify "good" architectural features. The SoS in those examples includes satellites with various internal configurations and operating in different orbits, and servicing satellites that can provide different types of support to extend the operational lifetime of the other satellites. Figure 8.3, from [86], shows the steps of this process. Existing architectures characterized by certain features are sorted according to user-defined SoS metrics. Neural Networks are then used to associate architectures that exhibit the desired performance (metrics can also be weighted) with their features, that is to identify architectural patterns that result in desirable behavior. This information can be used to design new architectures with a required type or level of performance.

Later, UQ techniques can be added to quantify how much each of the architectural features impacts each of the SoS metrics. Figure 8.4 illustrates how architectural features defined in [86] have different amounts of impact on the SoS-level metric of resilience of the architecture, as defined in [80].

In the second type of applications, AI is used as a fast, smart, and potentially learning data digging tool to provide input to the desired SoS tools. The complexity and size of typical SoS problems, as well as the uncommon form of the required data (for example, partly qualitative information about the impact of policies or stakeholder decisions) often cause difficulties in the gathering and sorting appropriate input for SoS tools. Advanced specialized agents can be designed and trained as learning scouts to find the necessary data in order to enhance SoS models and tools. For example, in a combined effort by NASA Marshall Space Flight Center, Purdue

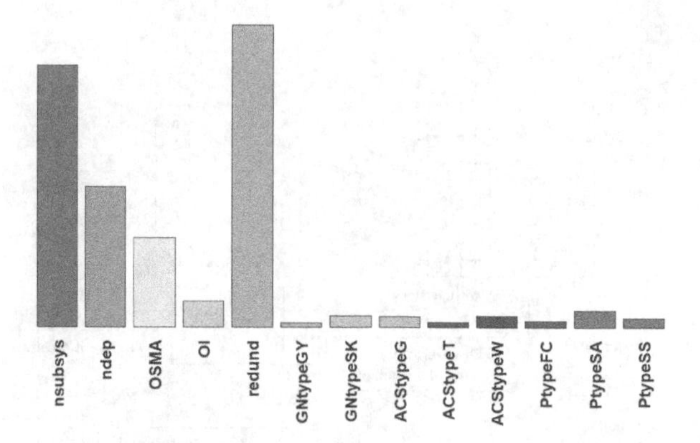

Figure 8.4: Relevance of various architectural features and patterns on the SoS-level metric of resilience, from [86]. The redundancy, number of on-board subsystems, and their dependencies are the most important characteristics driving resilience, followed by Orbital Semi-Major Axis (OSMA) and Orbital Inclination (OI). The other features, including various types of Guidance and Navigation systems, Attitude Control systems, and Propulsion types, have little impact on resilience.

University, and ai-one, Inc., AI agents have been developed and trained on a large technical repository to interpret plain language and score technical articles based on how much their content can provide specific information about Cryogenic Fluid Management technologies [84]. The outcome of this selection was then used to inform a model built in the Systems Developmental Dependency Analysis (SDDA) tool (Subsection 7.1.4) to provide decision-making on technology prioritization. Figure 8.5 illustrates the process of ai-one's AI agents training.

8.2 UNCERTAINTY

8.2.1 UNCERTAINTY IN SYSTEM OF SYSTEMS

Like all the topics in this chapter, uncertainty and its treatment is a vast topic. We shall only hope in this short section to provide insights on sources and implications of uncertainty in SoS. In fact, in many previous chapters we have discovered uncertainty, starting with the beginning of Chapter 1 and the concept of "deep uncertainty" explored in the LTPA of the RAND Corp. Then we found that the very essence of what distinguishes SoS (operational and managerial independence) makes obvious that the ability for one participant (or the SoS architect, should one exist) to know the actions and associated outcomes of another is and will always be imperfect. Add to this the externalities always present in the real world (e.g., weather in an air transportation SoS, demand fluctuations in a smart grid SoS, etc.) and there is much uncertainty to consider in SoS.

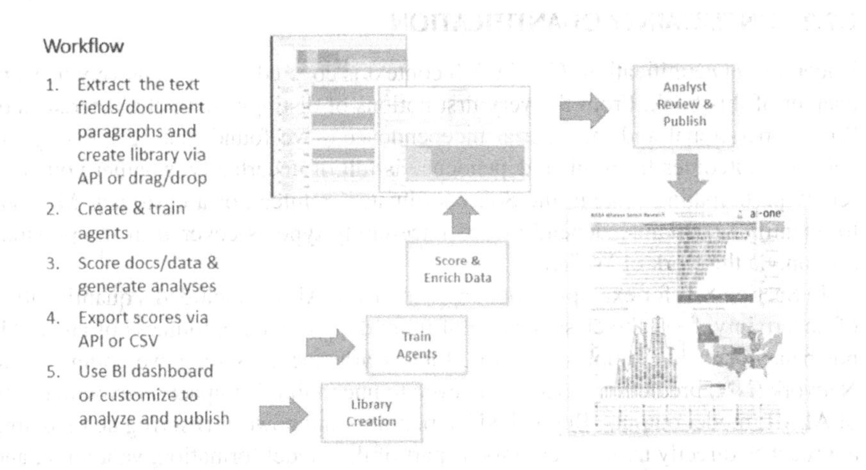

Figure 8.5: Phases of ai-one's AI agents training: implementation of database of abstracts; creation of agents; scoring of the abstracts with agents; presentation of results in a Business Intelligence dashboard.

SoS M&A must account for relevant uncertainties. But what is relevant? Ah, back to that question of how defining the problem is so hard, but important! Embedded in each of the three phases in the DAI process are specific elements that relate to uncertainty. One of the best frameworks for dealing with uncertainty in SoS is Walker et al. [168]. First, a simple typology is offered in which three levels of uncertainty exist between the extremes of "determinism" and "indeterminacy": statistical uncertainty, scenario uncertainty, recognized ignorance. The statistical type applies in settings where deviation from the true value can be characterized statistically; in SoS, we can imagine using historical data to provide a distribution for some resource behavior (e.g., a system capability as in our RPO examples in Subsection 7.1.1) or operational externality (e.g., localized, seasonal inclement weather and/or aviation fuel price as in air transportation examples). Scenario uncertainty recognizes the inability to validate the correctness of the many assumptions in making and using scenarios; in SoS, we can imagine scenarios in which the potential list of systems that may participate, in large part because they do not yet exist (e.g., supersonic transport aircraft as part of the commercial aviation)! Finally, recognized ignorance represents fundamental uncertainty about the mechanisms and functional relationships being studied; in SoS, one might image the uncertainty in how the effects of climate change would impact the air transportation system of 2040.

In addition to these three levels, Walker et al. [168] suggest that the location (within the problem definition) is an important and distinguishing dimension. By location they refer to whether the uncertainty resides in the model, the inputs to the model, or the context surrounding the model. This "location" analysis of uncertainty is consistent with the entity-centric abstraction displayed in Figure 3.4.

8.2.2 UNCERTAINTY QUANTIFICATION

Uncertainty Quantification (UQ) in SoS context is covered quite extensively in every chapter of this book. From the very first notions of distinguishing characteristics of SoS – operational and managerial independence – we found that deciphering the behavior/outcomes from this independence is full of uncertainty, whether you are a fellow participating system, the SoS coordinator/architect, or a customer. Methods to quantify uncertainty depend on the uncertainty type as covered in the previous section via the work of Walker.

In Section 8.1, for example, both applications of AI/ML included a quantification of uncertainty, from the classification of input data, to the randomness of agent/AI-bot behavior, to the employment of Monte Carlo methods for developing Neural Network (NN) prediction models, and even to uncertainty's impact on explainability of AI/ML model outputs. Beyond NN's, popular (and powerful) surrogate modeling approaches directly treat uncertainty as part of the model formation, validation, and use. The wide range of permutations of Gaussian Process models is a particular good example. In Section 8.3 on complexity, we will also see uncertainty plays a central role in the very definitions and conceptions of complexity, let alone the many measures and metrics that seek to quantify complexity.

The previous chapters covering ABM (Chapter 6) and Network Theory (Chapter 5) for modeling and a variety of SoS analysis tools for analysis in Chapter 7 have demonstrated the unique ways in which uncertainty is incorporated and assessed. Even the basic FSAs, the simplest form of an agent, may have probabilistic transitions that are driven in some way by uncertainty. For sure the most complex agents, such as a complicated, AI-enabled BKI-model agent, would have a much larger set of uncertainty sources, and this subsequent agent-based simulation will need to include uncertainty quantification, either through probabilistic sensitivity studies, Monte Carlo-type studies, or myriad other sampling/statistical analyses. Many applications in network theory, especially the modern network 'science' we cover in this book, directly operate on uncertainty, starting from assessing uncertainty in data when forming degree distributions and the underlying topology of real-world networks to the analysis of dynamics of, and on, networks. In very large SoS networks "operating in the wild," measurements of changes in the topology, especially on an individual/specific node and link basis, will almost never be deterministic. In fact, our experience applying network theory to air transportation confirmed this very point continuously; the ability to use network topology feature characterization to track changes over time, and to correlate these uncertain prediction models to performance metrics decision-makers care about, has been the absolute necessary mindset.

8.3 COMPLEXITY

8.3.1 CAN COMPLEXITY AID SOS M&A?

As described in Herbert Simon's classic book *Sciences of the Artificial* [149], and as evident from the Solberg Chart (Figure 1.1), there have been several waves of complexity theory that have influenced the systems analysis and design world. Complexity, with its ties to so many fields including mathematics, biology, ecology, and computer science, is for many an alluring concept so full of possibilities for uncovering elemental explanations for the amazing natural (and artificial) systems we see around us. Yet, harnessing this tremendous amount of science and insight to practical use in modeling, design and decision-making contexts has remained difficult across many fields of endeavor. This is true for SoSE. Perhaps a basic review of why we are interested in complexity will help.

In both natural and engineered systems, we *need* complexity to achieve higher levels of capability. A modern commercial airline aircraft is more complex than a General Aviation (GA) aircraft, and for good reason. An airliner needs to transport a large number of the most precious payload type – people – safely and (fairly) comfortably and do so every single time. Likewise, when first defining SoS in Chapter 1, we found the same motivation – we seek SoS for the capabilities it brings that are not possible by component systems alone or in isolation.

But how might an SoS architect view the matter and determine how much complexity is necessary? Figure 8.6, from [157], portrays a likely view such architect would take, seeking to identify design alternatives that achieve the required performance (with some margin, in most cases) with the minimal requisite complexity. These desirable solutions are indicated by the green solutions that lie on the complexity/performance frontier. The grey solutions do not meet the performance objective (with margin) and/or possess extremely high complexity. Finally, the red solutions... well, these are really bad choices! They achieve performance levels far above the requirement (gold-plating) and they do so with very high complexity, and with it the higher probability for unexpected failures.

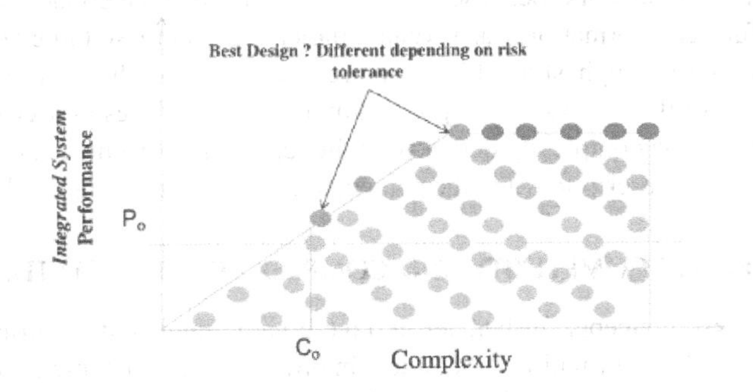

Figure 8.6: Complexity as means of characterizing the SoS design space and delineating architecture alternatives.

But how does one compute/quantify/measure this complexity in a manner that makes Figure 8.6 useful and authentic to a decision-maker? This too has proven quite difficult. Yet in attempting to tackle the task, we can learn much about the specific ways in which SoS, and the modeling and analysis of SoS, can benefit from complexity concepts. The following seemingly simple "quiz" further motivates the task at hand. Try the quiz shown in Figure 8.7.

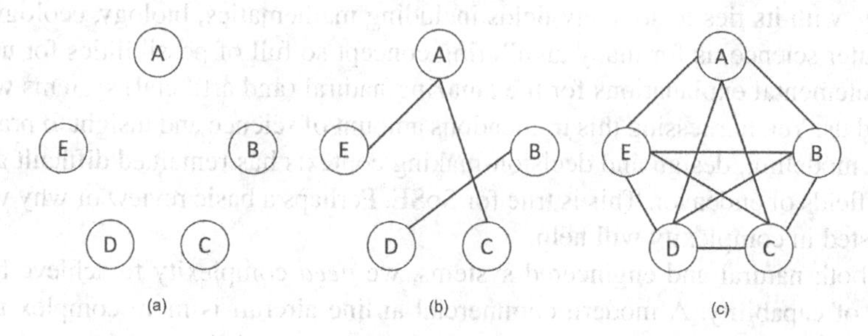

Figure 8.7: A simple quiz: Which of the three networks is more complex? Why?

In our experience, most students choose network (c) as most complex, with networks (a) and (b) competing for second place. The ensuing discussion is often interesting, sometimes contentious, but in the end illustrative of several practical lessons. First of all, some argue that network (a) does not qualify as a network since there are no edges. We generally leave this deep question to the graph theoreticians! Most of the voters for option (c) justify their answer because it has the most edges. Those voters who vote for (b) recognize that it is not a regular graph as networks (a) and (c) are. Given the limited context provided, literally just the depiction of the networks, this appears to be the most justified choice. For example, with equally compact description, we can describe to another person the essence of network (a) and (c). Using nomenclature from Chapter 5, we need only two parameters (number of nodes, *regular number*) to show that (a) is (5,0) and (c) is (5,4). Pretty simple. Network (b) would require more parameters, essentially upper triangle of the adjacency matrix for this undirected network, because of the heterogeneity of the connectivity.

Now, if further information was revealed indicating that these three networks are merely depicting the highest level in a hierarchical structure, then "all bets are off" and the details of the structure (and potentially behavior) of these lower levels must be provided. So, the complexity depends on the representation chosen, the degree of granularity of that description, and the perspective (and assumptions) of the observer.

8.3.2 EFFECTIVE COMPLEXITY AND COMPLEX ADAPTIVE SYSTEMS

To explore these concepts further, let us take a short intellectual journey through one pathway of the information-based view in order to arrive at some useful insights for SoS M&A. Guidance on this journey is taken in large part from the late Nobel Laureate Murray Gell-Mann and his book *The Quark and the Jaguar* [72]. When it comes to complexity, there is no better guide.

First, let us step back and reflect on our earlier notion of agents and agency and their treatment as a system. A kind of generic agent with some interesting behaviors is something we can term a Complex Adaptive System (CAS). What follows is an exploration of how we might characterize a CAS and distinguish its variants, inspired by Gell-Mann's presentation. In short, a CAS is an agent with a schema that interacts with its environment. To understand the complexity in a CAS, we will take an information theory viewpoint. And, by the way, at the end we will return to SoS, for a collection of "α-level CASs" is a β-level CAS; the rules of SoS compilation still apply.

Simplicity means "once folded," complexity means "braided together." We will see manifestation of this notion of braided together in the exploration of network dependency and coupling as a source of complexity. One of the simplest measures of complexity (or simplicity) is "the shortest possible time to do something" or "how long a message is required to describe the properties of a system." But we have to consider what level to describe the system, what degree of coarse-graining is at hand.

Crude Complexity is defined as the "length of shortest message that will describe a system, at a given level of coarse-graining, to a person at a distance using a language all understand" [72]. So, how can we go about quantifying the "shortest" message? Well, in information theory we have the notion of Algorithmic Information Content (AIC), defined as "the length of a shortest program to do something, then stop" [72]. This length can be measured by the number of bits in the program, where a bit is a binary choice between two equally probably events (0 and 1). But, in almost any case except degenerate ones, we will never know what the AIC is; we'll never know if a string of bits is truly the minimum one. This reminds us of Goedel's Theorem: "Given a system of axioms in mathematics, there will always be propositions that are true but undecidable based on these axioms." However, we can say something about the opposite extreme: the AIC is largest for random strings. Thus AIC may not be a useful measure of complexity for us; it does us little good to say the most complex thing we have is a random string.

So, if Crude Complexity and a simple use of AIC do not work for us, it seems more clear that a measure for Effective Complexity is not simply the AIC, but the "length of concise description of the regularities of a system" [72]. And this length is now dependent of the description, and a description can only be forged by an observer and this observer itself is a form of a CAS. So, a CAS observer is an agent that observes and then compresses to find regularities, and from these compressed regularities embodies a schema. As new data comes in, the schema must be reevaluated. Figure 8.8 displays this basic CAS architecture.

So, in final form, Effective Complexity is defined as: "relative to a CAS observing it, the length of a concise description of the regularities of a system identified in a schema" [72]. If the CAS observer is bad, effective complexity will be measured lower than it really is, a recipe for (negative) emergent behavior.

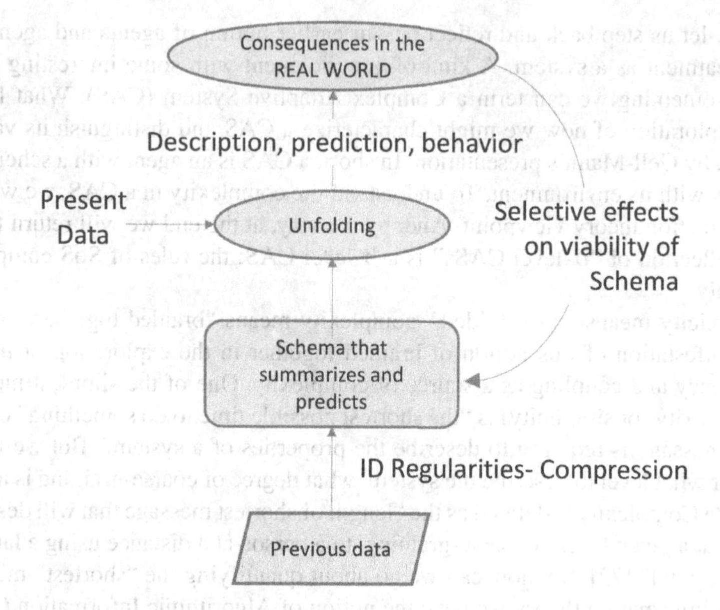

Figure 8.8: Basic architecture of a Complex Adaptive System (CAS), after Gell-Mann [72].

8.3.3 SOURCES OF COMPLEXITY

Now equipped with a better understanding of effective complexity, and the role of a CAS in computing it, it is appropriate to briefly list some of the potential sources of complexity especially in the SoS context. Many of these sources contribute to lengthening the description for regularities that can be identified in SoS.

The first source is the particular hierarchical description of SoS and the abstraction choices that are made in forming it. Figure 8.9, a simplified version of Figure 3.1, illustrates the delineation of networks in the hierarchy and the modeled interactions. Of course what is not shown are the unmodeled/unknown interactions that are missed by the representation. However, at least according to Herbert Simon [149], careful choices in modeling hierarchy and dynamics within it can be critical in understanding and designing effective complex systems. Simon pinpoints the crucial factor in designing what he calls "nearly decomposable systems." In such systems, only the most important interactions among levels of hierarchy that matter in the long run are maintained; gyrations in the short run are inconsequential and thus can be neglected (not part of the regularity-identifying schema). This is a quite deep concept that is also referred to and further explored in [72], especially from the quantum-mechanical viewpoint, and it is worthwhile reading for the SoS enthusiasts with theoretical/philosophical bent.

The second source, the size of SoS, is analogous to the third source, the heterogeneity of its components, in that both are features of the composition. Typically, the more components and the greater their characteristic diversity, the greater the

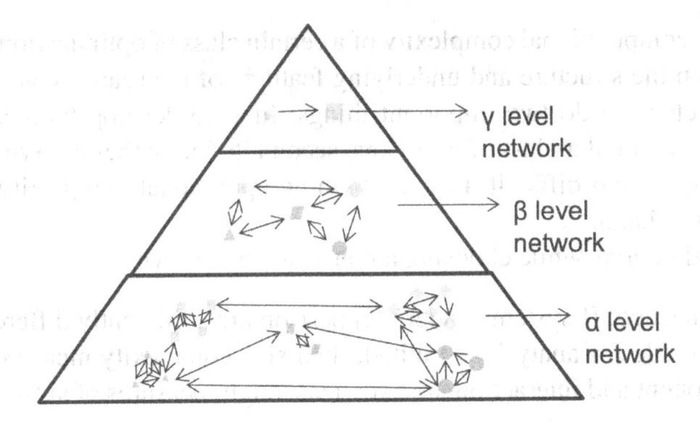

Figure 8.9: Interdependencies between and among SoS hierarchical levels is key source of complexity.

effective complexity. An example of this was seen in the quiz previously presented in Figure 8.7 where the heterogeneity in degree for the middle network indicated its greater complexity due to longer required description.

Fourth, as a dynamic, open system, SoS thus possesses the essential character of a complex system under the prevailing view of the physics community. With a porous boundary and a fluidity in terms of the systems that may enter, exit, or move within the existing hierarchy, SoS can generate much complexity because of these behaviors. Further, behaviors and changes in composition may proceed under multiple time-scales which in turn increases the likelihood of more unforeseen interdependencies. More broadly, the structure of the network topology itself becomes harder to predict and produce under such dynamism.

Finally, we highlight the multiplicity of perspectives among SoS participants, [140], which is a root cause of interoperability issues and a major source for SoS complexity. The operational and managerial independence of SoS components, essentially by definition, implies that these independent actors (agents) will look at the same facts and measurements a little bit differently. And thus they may misinterpret and form imprecise schemas. This in turn increases the complexity of the other participants and the length of their schemas, and so on.

8.3.4 COMPLEXITY METRICS

Complexity metrics are necessary for the goal of identifying design alternatives that have sufficient complexity to meet performance requirements (and reject known uncertainties) but no more, as shown in Figure 8.6. A prodigious amount and variety of complexity metrics are available in literature; however, the utility of their application depends on the problem under consideration.

There are information-based, computation-based, and structural-based views of complexity, among many others. There are certainly examples of quite significant utility in each of these views and their associated body of knowledge. For example, the optimization and operations research communities have developed powerful tools to establish the computational complexity of a certain class of optimization problems based largely on the structure and underlying features of the search space. This has allowed researchers to do two important things: first, to develop tailored solution algorithms for a particular class of problems, second, to know (before trying!) which problems would be too difficult, i.e., too high computational complexity, with the finite resources at hand.

Some considerations while choosing a metric are as follows:

1. Fidelity needed: If systems under comparison are sufficiently different and time available for analysis is limited, then size complexity measures such as component and interaction are excellent rough measures of system complexity.
2. For similar systems, topology aspects such as coupling and modularity also need to be considered.
3. For systems with software-hardware/dynamic interactions require special considerations.

The measure of complexity we employ will also depend on the context in which it is to be used. We could measure system complexity, dividing it into internal and external components. Whereas the former will include notions of hierarchy, connectivity, variety of linkages, and strength of interdependencies, the latter will consider ability to predict future states and multiple time-scales.

Alternatively, we can consider *computer science* or *algorithmic* view of complexity as a measure of the difficulty (number of possible operations) of finding an optimal solution. Several different features of a software can be used as a measure of complexity, such as lines of code, number of nested loops, etc.

An *information theoretic* view of complexity is the amount of information required to describe the system. An example of measure under this view is the Kolmogorov Complexity, where the complexity of a string is the length of its shortest description. Thus, under this measure, a random string such as 'abigkeplwtss' has a greater complexity than an ordered one such as 'abababababab.' The Information Entropy Complexity metric quantifies the degree of randomness or disorder in information content. This is measured using the following formula:

$$Complexity = \Lambda \times ln(\lambda) \tag{8.1}$$

Here, Λ is the total number of components and interactions, λ is the total number of unique components and interactions, and 'ln' indicates the natural logarithm. This equation measures increase in system complexity due to increase in number of components and interactions.

A *network theoretic* view of complexity states that "complexity is the degree of order in a network," and thus considers notions of "order and chaos" and the concept of phase transition. Thus, here engineering complexity could be a function of features of the network such as its size, topology, level of abstraction, representation, dynamics, etc. Let us continue to take a network theoretic view of complexity and discuss its sources next.

8.3.5 EXAMPLE OF AN SOS-RELEVANT COMPLEXITY METRIC

As noted in the previous section, the complexity of systems depends on different aspects such as level of abstraction, type of representation, heterogeneity, dynamics, etc. It further depends on size and topology, including the coupling and modularity. While size measures the number of components, coupling measures the feedback loops, and modularity measures the presence of a hierarchical structure. In the following, we present and discuss a coupling complexity metric developed and explored by Tamaskar et al. [157] and highlight how it can be applied to an SoS (or complex, hierarchic monolithic) system design problem.

As the name suggests, the coupling complexity metric is centered on quantifying multiple aspects of coupling within a system (including SoS) using a network representation. This includes local density of interconnections (essentially, identifying modules in the network), the number and size/length of feedback loops, all of which affect how much information one would need to characterize those features and use them to explain behavior – a means to compute effective complexity. The coupling complexity is measured by following these steps:

1. Develop a weighted network of system 'structural representation' → architecture
2. Redefine the weights to account for indirect coupling
3. List all the possible paths between all the node pairs in the network
4. Importance of a link in the structural network depends on the frequency with which the link occurs in this list. Calculate the new weights by multiplying the original weights with this frequency.
5. Compute coupling complexity (C_{cc}) by

$$C_{cc} = \sum_{s=1}^{c} j_s (\sum_{i=1}^{n} W_i)_s + \sum_{k=1}^{m} W_k \tag{8.2}$$

In Equation 8.2,

- W_i = weights of links of a cycle
- W_k = weights of links not belonging to a cycle
- j = size of the cycle
- c = number of cycles
- m = number of links not participating in any cycles

8.4 MODEL-BASED SYSTEMS OF SYSTEMS ENGINEERING

Model-Based Systems Engineering (MBSE), according to the definition by INCOSE, is "the formalized application of modeling to support system requirements, design, analysis, verification and validation activities, beginning in the conceptual design phase and continuing throughout development and later life cycle phases" [66]. In particular, MBSE is a Systems Engineering methodology that focuses on creating and exploiting domain models as the primary means of information exchange, rather than relying on document-based information exchange.

The need for representations of components of SoS and their dependencies has been highlighted in detail in the description of the abstraction phase of the DAI process (Section 3.4). This is often accomplished through the use of networks (Chapter 5), which provide an effective way to model and analyze interactions between systems in an SoS, as well as to visually represent the topology of an SoS architecture. However, MBSE goes beyond simple network representations of systems and their interdependencies, since it provides support to multiple engineering activities.

With regard to Systems Engineering, many practical approaches to implementation of MBSE have been developed. The most common and widespread of these is based on the use of the Systems Modeling Language (SysML, [1]), created in 2003 as a dialect of the Unified Modeling Language (UML) to provide system architecture modeling, including formal syntax and semantics of modeling diagrams. Another tool commonly used to carry out MBSE is the Department of Defense Architecture Framework (DoDAF) [52], which provides views to organize and visualize complex systems architectures. A study by Giachetti [73] evaluated how to utilize DoDAF to implement MBSE.

However, as explained in Chapters 1 and 2, System of Systems are special types of systems, and much attention must be given to the fundamental characteristics of SoS when applying MBSE. For example, the user needs to select appropriate representations of the dependencies between systems, of the interfaces that must be designed to deal with the operational and managerial independence of the component systems, and so on. Some authors have described application of specific SysML or DoDAF methodologies, diagrams, and views to SoS problems. For example, Guariniello et al. [82] describe which SysML diagrams and DoDAF views can be used to execute various steps of the SoS modeling and analysis tools described in Chapter 7. Figure 8.10 shows the association of MBSE tools and steps of execution of analysis with the SODA tool.

Similarly, Huynh and Osmundson [103] linked SoS architecture, DoDAF products, and SysML language, with examples in the naval field. The MITRE corporation developed SoS case studies with a SysML executable model. However, a much more thorough study of the use of MBSE for SoS modeling and analysis was performed by Kevin Bonanne [28], who explored application of MBSE to directed, acknowledged, and collaborative SoS, and detailed how SysML products can be used to address aspects of SoS. This is a summary of some of the conclusions of Bonanne's work:

1. Practitioners define managerial independence in SysML diagrams for SoS
2. Operational independence of component systems is specified by allocation of activities to system blocks
3. Evolution of the SoS comes from modeling of separate stable intermediate forms
4. Emergent properties come from system interactions depicted in the models
5. Geographic distribution comes from specifying location properties
6. Heterogeneity comes from customizing blocks representing systems
7. Trans-domain nature comes from the set of systems modeled and the goal of the model
8. Networks come from the use of flow ports and connections

SODA steps	DoDAF views	SysML diagrams	Available information
Problem definition	AV-1: Overall goals and objectives		• provide problem definition • identify systems involved • identify capabilities that the user wants to analyze / quantify • identify possible flexibility / system replacement
	OV-1: High-level graphical/textual description of the operational concept	Block Definition Diagram Package Diagram	
	SvcV-3: Relationship among or between systems and services in a given architecture	Activity Diagram	
	SV-4: Functions performed by each system	Activity Diagram	
Operational dependencies	AV-1: Overall goals and objectives		• identify possible SODA topologies, based on overall goal and objectives, and desired analysis
	SV-1: Identification of systems and their interconnections	Block Definition Diagram Package Diagram	
	OV-2: Resource flow exchange OV-5a: Operational Activities Decomposition	Internal Block Diagram Block Definition Diagram	
	CV-4: Capability Dependencies	Block Definition Diagram Activity Diagram	
Self-effectiveness and operability	AV-1: Overall goals and objectives		• identify the meaning of SE and operability, based on goal and objectives, and functions and services performed by the systems
	SvcV-3: Relationship among or between systems and services in a given architecture	Activity Diagram	
	SV-4: Functions performed by each system	Activity Diagram	
Parameters of the model	AV-1: Overall goals and objectives		• identify the parameters representing the dependencies in a SODA model
	SV-1: Identification of systems and their interconnections	Block Definition Diagram Package Diagram	
	SV-7: Systems Measures Matrix	Parametric Diagram	
Analysis	SV-7: Systems Measures Matrix	Parametric Diagram	• identify values / probability density functions for SE • perform the required SODA analysis
Synthesis		Activity Diagram	• identify how the network can be modified, based on requirements, flexibility rules, and results of analysis
		Requirements Diagram	
		Sequence Diagram	

Figure 8.10: Associating MBSE tools and SODA tool [82].

It is evident from this (non-exhaustive) list that some of the basic MBSE concepts and their implementation in SysML need to be adapted to explicitly address features of SoS. A correct execution of these steps provides a powerful tool to model and analyze SoS and to provide a common source of truth to all the involved stakeholders.

8.5 CHAPTER SUMMARY

SoS M&A can be enhanced by both existing and new/emerging representation, analysis and synthesis techniques. Four prominent examples of this are presented in this chapter: AI/ML, Uncertainty Quantification (UQ), Complexity Theory, and Model-based Systems Engineering (MBSE). Each section treating these four examples includes both an introduction to the area as well as specific examples of application to SoS that point to the potential value motivating their adoption. Along the way, a number of potentially revolutionary developments are identified.

Part III

Examples of Application of System of Systems Modeling and Analysis

9 Advanced Air Transportation System of Systems

We have been using the air transportation System of Systems as a running example throughout the text. In this chapter, we will extend those ideas to the modeling and analysis of the same system to figure out its future evolution. Numerous technological advancements such as novel aircraft configurations, electrification, and new propulsion system concepts, and factors such as the emergence of new markets, including use of unmanned systems for cargo delivery and passenger transportation within urban environments are collectively leading to revolutionary changes in air transportation. We will introduce the problems facing engineers of such systems and discuss how the three-phase DAI process can be successfully employed to guide its eventual and inevitable introduction.

9.1 PROBLEM INTRODUCTION

Air transportation is an excellent example of the type of system which can be categorized as an SoS. It involves a wide diversity of constituent systems most of which have long development and operation times. Their life cycle costs are high which is why investment decisions have a long-term impact on a stakeholder's budget. The stakeholders too are numerous and each has their own objective for participation. This is complicated by the global nature of this industry which adds layers of complexity in the form of international policies and regulations on top of the already complex operational problems.

Despite this enormous complexity, the present-day aviation is both an example of an industry which successfully integrates cutting-edge technologies with standardized operational procedures across national borders and also witnessing rapid investment in development of new technologies and markets. All stakeholders of this industry have a natural incentive to speed up incorporation of these new technologies plus the accompanying operational paradigms to increase profits, improve safety, and enable novel markets such as those of transportation within urban environments. Needless to say, such rapid development is an exciting opportunity for us to use an SoS approach to the evolution of this industry.

First, we need to recognize that regardless of the speed of development of new technologies, their incorporation into operation will be a slow process. To maintain the safety record of the aviation industry, each technological or operational innovation will need to be proved safe even as it integrates with legacy systems.

We will need to account for multiple layers of system hierarchy, complexity of the networks and heterogeneity in constituent systems, their operations, and guiding policies.

9.2 DEFINITION PHASE

9.2.1 OPERATIONAL CONTEXT, STATUS QUO, AND BARRIERS

The aviation industry is poised for rapid and revolutionary changes in all aspects from new technologies to operational paradigms. The introduction of new technologies such as distributed electric propulsion is enabling new markets such as those that serve urban and regional areas. However, our current experiences with traditional aviation do not carry over neatly to these new markets due to several reasons. The present nature of the air transportation system is the result of decades of gradual technological and operational advancement. Tube-and-wing configuration aircraft are the staple of all markets in commercial aviation. Rotary-wing aircraft – especially helicopters – fill in for most of the rest of the commercial aviation market. Operations are commanded by voice-dependent communication with air traffic controllers. Nearly all operations happen at large public airports which are frequently located away from city centers. The large size and operating economics of current aircraft also limit where and how they can be profitably utilized.

Contrast the above status quo with the expected changes with new technologies. First, advancement in electric and hybrid-electric propulsion technologies is giving impetus to new use cases for commercial aviation. The vehicles used for operations in some of these markets are expected to be smaller than traditional aircraft; aircraft being developed for urban air operations, for example, are expected to fall in the range of one to five passengers [77]. The rate of operations is also expected to be higher, with thousands of aircraft likely to be simultaneously airborne in the mature state of urban air operations.

Advanced Air Mobility (AAM) is a concept which envisages an integrated aerial transportation system supporting numerous different types of vehicles both in urban and regional markets. These new vehicles and their operations are expected to integrate seamlessly with the National Airspace System (NAS), thereby extending the traditional long-distance aviation markets to shorter regional and urban use-cases. Examples of the new markets forming under AAM include those of delivery of goods in urban environments via drones and passenger and cargo operations linking urban and rural areas.

One of the features of such an undertaking is that we need to keep a broad focus beyond just the vehicles and include considerations of airspace, infrastructure, and community involved (Figure 9.1). The number and variety of stakeholders involved and the complexity of the system mean that development of technology and its deployment will be phased and iterative; coordination between all stakeholders supported by decision support tools, of which systems of systems modeling and simulation are an essential component, will be necessary [35].

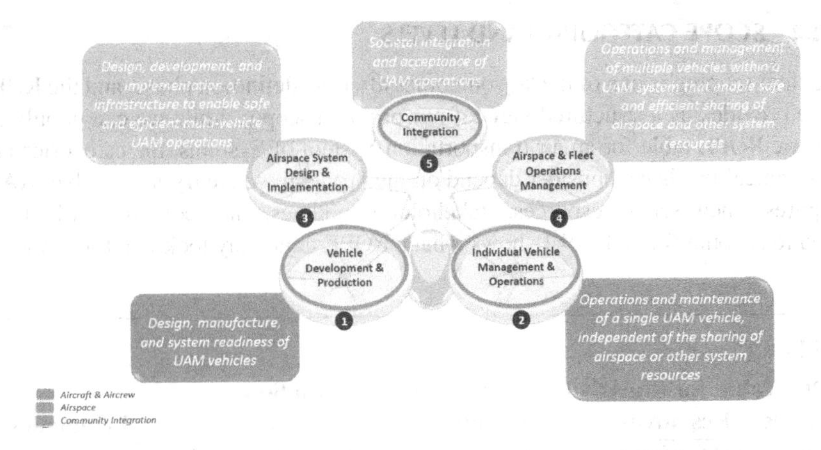

Figure 9.1: Five pillars of the UAM framework [90].

The enabling technologies are novel and yet need to be integrated with the existing systems and operating paradigms. New vehicle configurations and propulsion systems will require new testing and simulation capabilities. These new vehicles will be supported by new ground, air, and space infrastructure and new air transportation management systems, which means that early insights into this industry's concepts of operations will help us make the right investment choices for technology development. Future operations are expected include higher levels of autonomy and the air traffic control may provide only supervisory role rather than active clearances. This means that cybersecurity will need to be an inherent consideration given the increasingly higher levels of autonomy of operation of all systems.

The Committee on Enhancing Air Mobility [35] lists safety, security, social acceptance, resilience, environmental impacts, regulation, scalability, and flexibility as factors that will determine the success or failure of AAM. The new markets enabled under AAM will operate in much more proximity to people than traditional aviation. Thus, for these markets, community acceptance will play a key role in its eventual success or failure. Environmental factors such as the impact of noise and visual pollution will carry higher penalties than those for traditional aviation markets. Even the weather will have greater impact, since the small size of the aircraft will make them more susceptible to wind gusts. Finally, there currently exists no infrastructure to support this new market and everything from "vertiports" to airspace services infrastructure will have to be designed and built from scratch. New infrastructure will require high investment which will rely on industry willingness to invest and take risks.

Due to the cutting-edge nature of technologies involved and the diversity of stakeholders affected, we need to identify potential challenges early in the design stage and take corrective design decisions to reduce development time, costs, and risks involved. The costs and benefits of AAM need to be understood early as new stakeholders and use-cases will continue to arise even during deployment of supporting technologies.

9.2.2 SCOPE CATEGORIES AND LEVELS

Problem scoping is one of the key outcomes after the definition phase, and the ROPE table provides us a structured view of our problem scope. While we can populate a generic ROPE table for an air transportation System of Systems, the categories and hierarchical levels we consider depend on the problem we are trying to solve. AAM requires a new set of resources, stakeholders, policies, and economics, which we need to account for. Table 9.1 shows what a ROPE table may look like for AAM.

Table 9.1
ROPE Table for the Advanced Air Transportation SoS

Levels	Resources	Operations	Policies	Economics
α	VTOL aircraft, vertiports, charging stations	Flight mission profiles from vertiports	Rule for aircraft registration, operation	Operating costs of aircraft, vertiports
β	Operator (airline) fleets	Managing and operating fleets	Policies for operating fleets	Economics of fleet operator
γ	Network of operators, infrastructure	Operations in geographically constrained environments	Regulations of low-altitude operations over population centers	Economics of on-demand transportation
δ	Networks of resources from different AAM markets (UTM / UAM / RAM)	Integrated operations of different AAM markets	Policies for operation of vehicle classes in different airspace classes	Economics of all AAM markets

α-level resources include entirely new types of aerial vehicles. Because these vehicles are required to operated in geographically constrained environments, they are expected to be aircraft capable of vertical takeoff and landing (VTOL) and make use of either fully electric or hybrid-electric propulsion systems, which means that they will require development of supporting charging infrastructure. Passenger aircraft serving the urban air mobility market will operate from vertiports. The other categories at α-level also relate to individual aircraft or drone, with the operations category relating to operating procedures (mission profiles), the policies category addressing the rules governing certification and operation, and the economics category relating to their operating costs. β-level categories include resources, operations, policies, and economics of a single fleet operator (the equivalent of an airline in traditional aviation). At the γ-level are the categories for each of the AAM markets separately, while the δ-level categories are for all AAM markets combined. The different AAM markets include those of drone operations (UAS traffic management, UTM), urban air mobility (UAM), and regional air mobility (RAM).

While the main focus of this chapter is on advanced air transportation, the new aviation markets will integrate in the existing National Airspace System (NAS). As a demonstration of the 3-phase DAI process to SoS modeling and analysis, it is therefore instructive to also include discussion of some of the SoS problems from traditional aviation. We will look at this discussion in form of a series of examples below. Each of these three examples considers a different version of the ROPE table for traditional aviation, shown in Table 9.2. Then, every time we study a problem in the air transportation SoS context, we can customize this table to suit our purposes.

Table 9.2
ROPE Table for the Air Transportation SoS; ICAO: International Civil Aviation Organization; WTO: World Trade Organization

Levels	Resources	Operations	Policies	Economics
α	Vehicles and Infrastructure	Operating a resource	Policies for single resource use	Economics of a resource
β	Collection of resources for a common function	Operating resource networks for common functions	Policies concerning multi vehicle use	Economics of resource networks
γ	Resources in a transport sector	Operating a collection of resource networks	Policies concerning sectors using multiple vehicles	Economics of a business sector
δ	National Air Transportation System	National passenger and cargo flight movements	National air transportation system policies	Forecasts of national air transportation market
ε	Global Air Transportation System	Global passenger and cargo flight movements	Bilateral agreements, ICAO regulations	WTO, global marketplace

9.2.2.1 Example 1: Robust, Scalable Transportation System Concept

Consider the problem of developing a high-level, integrated transportation concept (not an aircraft!), representing a view toward a new National Airspace System (NAS). We wish to know how to transform the system toward the use of novel technologies to achieve scalability in throughput, accessibility, and robustness. It is a design decision-making problem, so Engineering Design has a central role. The design intent is a service – transportation – but there is no central design authority. Yet, we wish to compare some designed model to a target outcome (e.g., if a future system can meet a particular demand). This "differencing," however, is ill-advised as it is a point design (e.g., designing to a certain demand level), and it misses many dynamics, variables, and possible inputs. It is not robust to uncertainty or to uncertainty over time. In short, we seek a systems of systems model generated using Network Theory and Agent-Based Modeling to "evolve" new NAS, measure goals, and uncover rules of behavior and network patterns that lead to scalable NAS. The ROPE table for this model will span all four categories, but we only need to consider levels β, γ, and δ in Table 9.2.

Figure 9.2 shows a schematic of how this concept can be modeled, including mobility and capacity networks and models of stakeholder agents. In the abstraction phase, we will identify the different resources and stakeholders involved along with their networks of interactions, and in the implementation phase we will develop models of these agents.

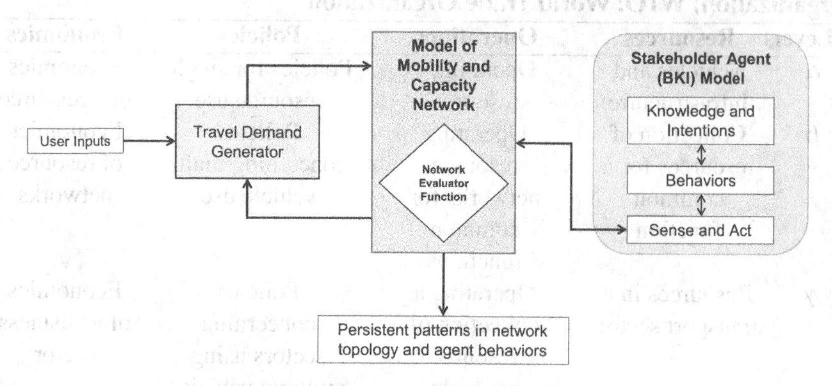

Figure 9.2: Schematic of transportation system concept study approach.

9.2.2.2 Example 2: Assessing New Technologies on Future Fleet and Emissions

In this example, we look at the impact that new technology has on aircraft performance, emissions, and operating cost – at the fleet level. Our objective is to develop a model of the airlines' decision-making on how they choose the composition of their aircraft fleet and how they use those aircraft, i.e., what are the decisions on aircraft allocation to different routes. This is the example we discussed in Chapter 3. For this problem, we consider the levels α through δ for all four categories in the ROPE table, as shown in Figure 3.7.

9.2.2.3 Example 3: Air Transportation Network Restructuring

Our first example developed a model of a transportation system concept and second example studied an airline's fleet composition and allocation decisions. In this problem, our task is to study how the transportation network itself evolves. Thus, our need is to generate more "complete" forecasts from a network evolution perspective, to improve fidelity of system-wide evaluations, such as those in the first two examples. We assume that we begin with a given fleet, which means that we need not consider the "Resources" category of the ROPE table. Network evolution results from the airlines' route planning decisions, which related to the "Operations" category of the ROPE table; additionally, we consider the "Policy" category at β, γ, and δ levels.

9.3 ABSTRACTION PHASE

In the second phase of our process, we draw out the constituents of our model – the resources and stakeholders, their networks of interaction, and the drivers and disruptors. These are the pieces with which we implement the model in the final phase of our process.

9.3.1 RESOURCES, STAKEHOLDERS, AND NETWORKS

Table 9.2 lists the resources at five levels of hierarchy of a generic air transportation system; these include the vehicles, especially aircraft, and individual pieces of infrastructure at the α-level, followed by successively larger networks of these resources. The stakeholders within an Air Transportation System (ATS) are many and diversified. They include the passengers who use the transportation service, airlines who own and operate the aircraft to meet their objective of profit maximization, airports and other infrastructure providers who are independent businesses with their own profit-maximizing interest, and regulators who enforce policies to ensure safe operations.

The resources and stakeholders of the AAM market fall into similar categories, though the list is longer and more varied (Table 9.1). For example, while most vehicles in traditional aviation are either fixed-wing or rotary-wing aircraft, the diversity of vehicle configurations in AAM is much wider. A passenger-carrying aircraft being designed for UAM includes configurations such as separate lift and cruise propulsion, tiltwings, tiltrotors, among others [159]. The stakeholders are equally diverse and include vehicle manufacturers, some of them with little or no previous experience, the public which is impacted by flight operations, and even new service providers.

Each of the resources and stakeholders can be organized in its own networks. For example, a provider of services to UAM (PSU) provides ATM services to UAM operations, and all PSUs are connected in a network which enables inter-PSU communication. Even the route networks are newly designed keeping in mind the restrictions and safety concerns of operations at lower altitudes in areas with high population densities. UAM operations, for example, will operate within a UAM Operating Environment (UOE), which is a flexible airspace area encompassing the areas of high UAM flight activity [95].

9.3.2 DRIVERS AND DISRUPTORS

What are the drivers of this system? Let us look at the drivers of traditional aviation first. Generally, countries with higher standard of living have more flights per capita. Thus, the economy is a strong driver of the air transportation industry. Similarly, policies and operating procedures favorable to safe and efficient flights are catalysts for the growth of aviation, which is one of the purposes of the existence of bodies such as the International Civil Aviation Organization (ICAO). Each of these factors also works in favor of AAM, but additional drivers play a key role.

UAM operations will provide on-demand transportation services within and around urban areas at speeds greater than current ground-based transportation systems [159]. This translates to time savings and increased economic value to passengers. Besides passenger transport, new aviation markets will also provide benefits to users such as security agencies, emergency response services, and cargo delivery. A range of vehicle platforms spanning from small remotely operated drones to larger unmanned aircraft can provide regional as well as "middle-mile" and "last-mile" cargo delivery services [35].

While the benefits are many, the challenges (and eventual disruptors) to AAM are equally numerous. Factors such as adverse weather, geopolitical issues, or poor economic conditions all provide headwinds to traditional aviation. Additionally, AAM faces new safety, security, regulation and certification, environmental, and community acceptance barriers [35, 159]. The desire to implement higher levels of autonomy will require development of new software and hardware for flight control, communication, and navigation. Cybersecurity will be a key consideration in all aspects of autonomous system design. Supplementary services, including weather forecasting and systems for collecting and sharing flight data, will need to be developed and certified alongside other infrastructure [35].

Figure 9.3 shows an entity-centric abstraction of AAM with a subset of resources, stakeholders, drivers, and disruptors. Let us now take the example problems of traditional aviation from the previous section through the abstraction phase.

9.3.2.1 Example 1: Robust, Scalable Transportation System Concept

When defining this problem, we recognized that both ABM and Network Theory are used in developing models of our concept transportation SoS. In the implementation phase, we will develop agent models of two stakeholders. In the current phase, we wish to develop networks of resources. For example, we can identify the network of routes connecting airports, where the nodes represent commercial airports and links represent flight segments between those airports. Using historical data available from sources such as Bureau of Transportation Statistics (BTS), we can assign properties to network nodes and links. For example, the different colors can represent the sizes of metropolitan areas (airports). Similarly, the amount of traffic on a given route can be the weight assigned to the corresponding link.

The outcome from our process will not be a network design, rather identification of possible paths of future network evolution under different scenarios. Hence,

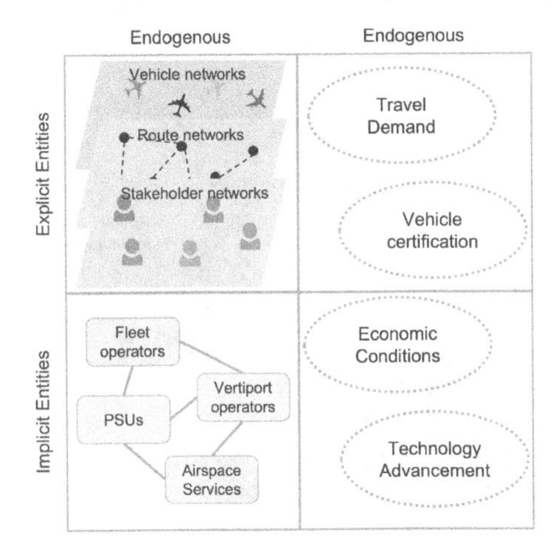

Figure 9.3: Entity-centric abstraction of advanced air transportation markets.

definition of future scenarios with respect to technology evolution and policy and economic changes will form part of the model's inputs. Any of these factors could be drivers or disruptors, depending on their impact on the overall system. Figure 9.4 shows an entity-centric abstraction of the model.

9.3.2.2 Example 2: Assessing New Technologies on Future Fleet and Emissions

There is one key stakeholder – the airline. Several abstraction layers for an airline decision-making problem were shown in Figure 3.9. We identify the set of aircraft owned by the airline. For this, a survey of historical fleet sizes is done, and the sum of all aircraft is divided into a set of predetermined size and technology-age categories. The number of aircraft in each of these categories is scaled to represent the size of the airline. Route network is determined similar to that in the previous example, i.e., by using available data from BTS.

Since airline decisions are impacted by market and economic conditions, we use available data on factors such as projected demand and economic growth as inputs to the airline's decision model. These are either the drivers or disruptors depending on the impact they have. Environmental factors are generally an output from the model, though we can impose environmental policy as a driver, for example by adding a constraint on the amount of emissions allowance that an airline is permitted.

9.3.2.3 Example 3: Air Transportation Network Restructuring

In this project, the objective is to forecast the evolution of service network topology. Changes to topology will be accompanied by corresponding changes in the dynamics including the development of new business models, policies, and technologies. To conduct system-wide evaluations, we input the current state of network, market, and

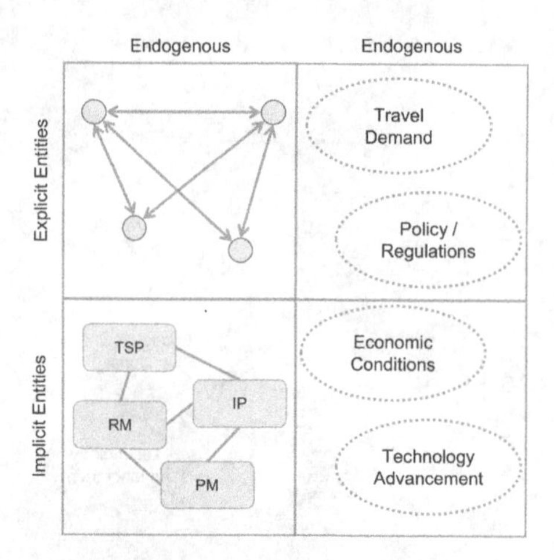

Figure 9.4: Entities in the model of transportation system concept. The resource network is the mobility and capacity network. Stakeholders include transportation service provider (TSP), infrastructure provider (IP), resource manufacturer (RM), and policy-maker or regulator (RM). The four entities on the right of this figure can be either drivers or disruptors depending on their impact.

policies, and run simulations predicting future states of operations, environmental impacts, etc. Abstraction does involve determining the current network topology and associated operations. Similar to previous examples, we can use available aviation market data to develop our models.

Once we have historical data, a challenge we face is identification of future scenarios of policy changes, concepts of operations, or even economic and environmental factors. Using the two inputs of historical data and future operational concepts, our implementation will produce the likely behaviors as measured by metrics such as levels of emissions, congestion, delay, etc. Figure 9.5 shows the models we need to develop for this problem.

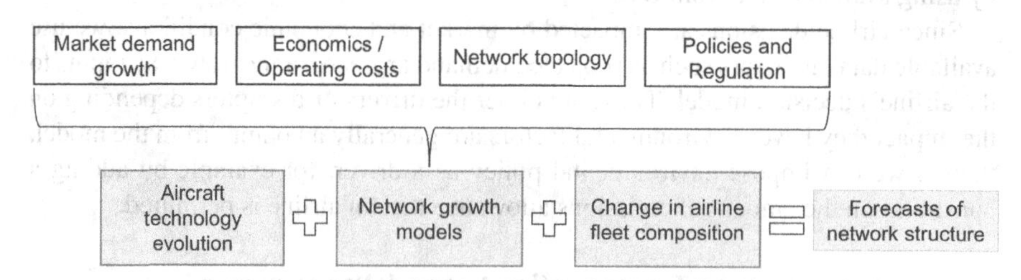

Figure 9.5: Abstraction for network evolution problem.

9.4 IMPLEMENTATION PHASE

In the final phase of our process, we begin to implement the models we have developed so far to test our hypotheses and guide decision-making. We can implement each resource and stakeholder as a separate object and link them in a higher-level model of network of interacting agents. Each of these objects will have its own goals, preferences, and decision-making models implemented as local methods. Separately, we develop algorithms to simulate coordination among agents and the resulting operations in form of system-level dynamics.

To demonstrate the implementation phase for modeling and simulation of AAM, we will use the UAM Vision Concept of Operations (ConOps) [95]. The ConOps provides a concept for further exploration of ideas for advancement and integration of UAM in the national airspace system. The ConOps [95] provides a description of UAM operations at the intermediate state of maturity (UAM Maturity Level - 4), which is a state with hundreds of simultaneous operations, closely spaced high throughput aerodromes (vertiports), many UTM-inspired ATM services, simplified aircraft operations, and low-visibility operations. These operations are enabled by a network of stakeholders some of which include the aircraft fleet operators, the providers of services of UAM (PSUs), the FAA, and Supplemental Data Service Providers (SDSPs). We will develop agent models of two of these stakeholders – the fleet operator and the PSU.

Figure 9.6 shows the BKI agent models of a fleet operator (Figure 9.6a) and a provider of services for UAM (PSU) (Figure 9.6b). The fleet operator owns, manages, and operates an aircraft fleet. In effect, the fleet operator provides transportation services to the end users, which are the passengers or cargo. The responsibilities of a fleet operator can be set as its behaviors in an agent representation; these include managing aircraft operations, meeting all regulatory requirements, and planning and execution of flights, among others (see the UAM Vision ConOps document for more details of the operator's responsibilities [95]). At the same time, the operator's goals (which become the intentions) include maximizing safety and generating profits. Note that this is a simplified representation, and a real fleet operator will have many more responsibilities and goals. During operation, an operator interacts with a PSU, the pilot of each of its aircraft, the passengers, regulatory agencies (e.g., FAA), and other stakeholders.

The PSU is responsible for providing ATM services to the operators. In brief, its responsibilities include supporting the operational planning of operators, ensuring that all relevant flight and network information is distributed to the rest of the PSU network, and providing data exchange services between operators and PSU network. A PSU is interested in increasing safety of operations and the throughput (of flights) in the network.

To implement the above agent models computationally, we define separate classes for the operator and the PSU. Figure 9.7 shows the class diagram (from the Unified Modeling Language) of both agents. The attributes of the fleet operator class (Figure 9.7a) include its knowledge of its aircraft fleet, the PSU it interacts with, etc. Figure 9.7a shows some of the methods of this agent class (which are the functions

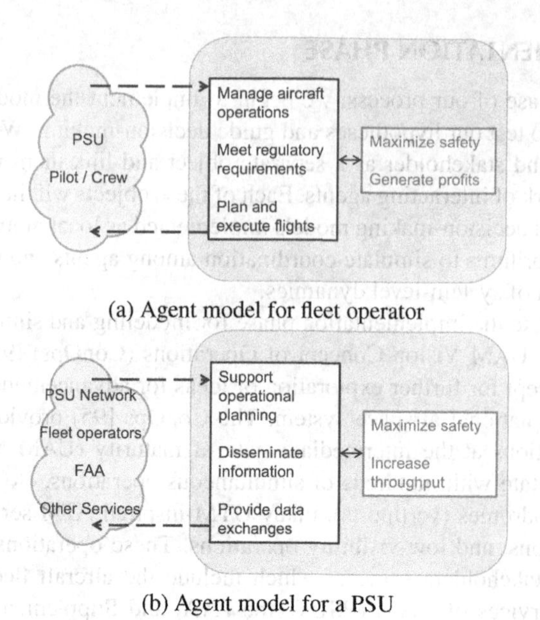

(a) Agent model for fleet operator

(b) Agent model for a PSU

Figure 9.6: BKI agent models for fleet operator and provider of services for UAM (PSU).

of this agent) including filing operations plans with the PSU, verifying the passenger manifest, and dispatching an operation for takeoff. Figure 9.7b shows the class diagram of a PSU, whose attributes include its knowledge of its fleet operators, the PSU network (i.e., the other PSUs it interacts with), etc. A PSU provides the functions of strategic deconfliction and negotiation for resolving potential conflicts.

To fully implement all agents to be able to simulate UAM operations, we need to define the stakeholders' agent models during all flight phases. The UAM Vision ConOps [95] also describes the responsibilities of all major stakeholders during gate-to-gate flight operations. Taking the example of a fleet operator, we can define one agent class with all functions and then define state transition rules so that the agent selects different tasks relevant to the current operational phase. This will be more clear from Figure 9.8 which shows the state transition diagram for a fleet operator. Figure 9.7a shows the agent state during the preflight phase of operations, which is the starting state for this agent. Thereafter, this agent takes on different responsibilities in each successive flight phase. For example, in the taxi and takeoff phase, the operator has the functions of approving taxiing and takeoff, and similarly for the rest of the phases.

Having defined agents for all stakeholders, we can set up a simulation of gate-to-gate operations by linking the agents into a communication network of interactions and observe the resultant UAM operations. Figure 9.9 (from [62]) shows a notional architecture of UAM operations as a network of stakeholder interactions. Finally, by defining different use-cases and scenarios, we can explore behaviors of all agents and

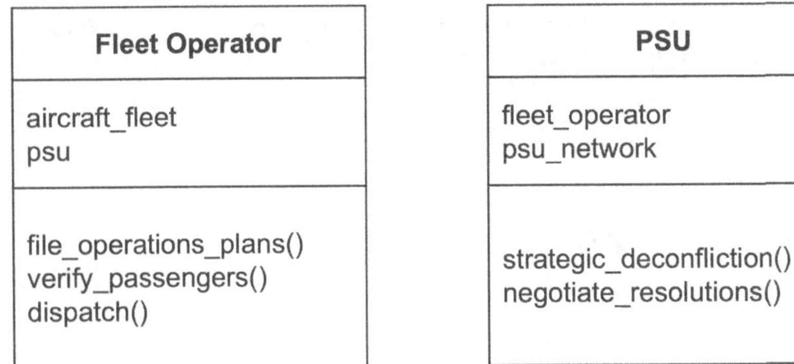

(a) Class diagram for fleet operator

(b) Class diagram for a PSU

Figure 9.7: Class diagrams for fleet operator and provider of services for UAM (PSU) showing names, attributes, and methods for classes of both agents.

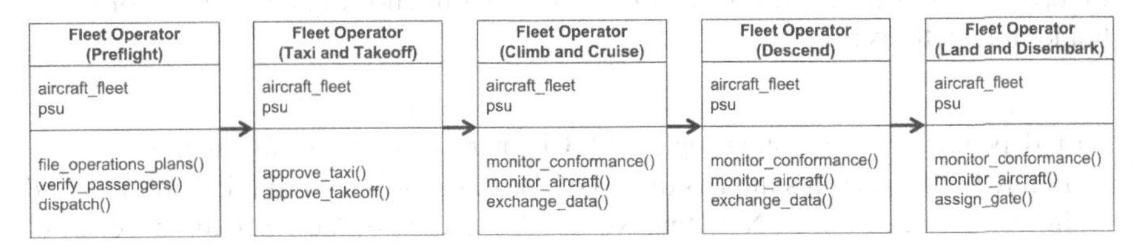

Figure 9.8: State transition diagram for fleet operator agent.

of the complete system; see, for example [35, 159, 164], for additional descriptions of UAM implementation and simulation studies.

In the following examples, we will see how each of the pieces of the model can be implemented, together with supporting models of policy and economic factors as inputs.

9.4.1 EXAMPLE 1: ROBUST, SCALABLE TRANSPORTATION SYSTEM CONCEPT

When defining this problem, we recognized that both ABM and Network Theory are used in developing models of our concept transportation SoS. We also stated that we do not wish to "design" the new system directly, rather examine its possible evolution over a range of scenarios. While factors such as economy, regulations, and technological advances are part of scenario definition and are inputs set by the modeler, the entire system is dynamic in nature and emergent patterns in network topology and system-level behavior also form part of the scenario. We use Network Theory (Chapter 5) to identify patterns in network topologies. Now, we want to find the patterns in rules of behavior (in stakeholders – travelers, service providers, infrastructure providers) that tend to produce these good network patterns, while accounting

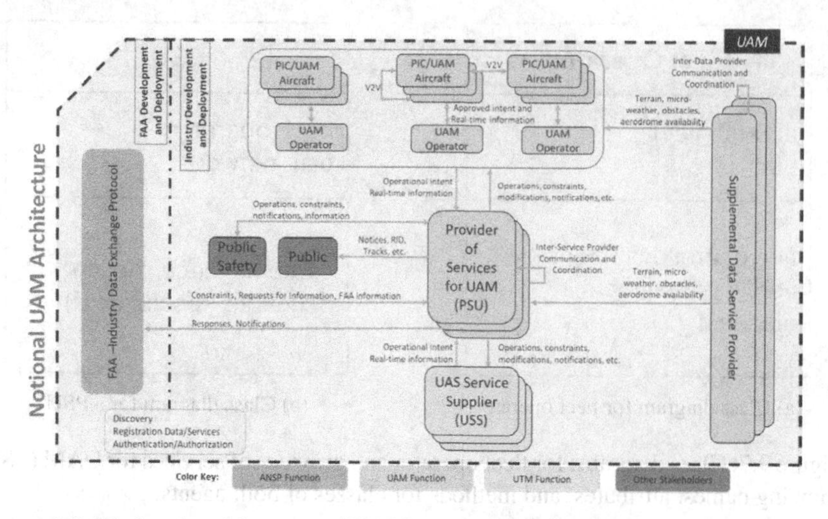

Figure 9.9: Notional architecture of UAM operations showing the network of stakeholder interactions [62].

for independence of action. Using ABM (Chapter 6) allows us this capability.

Figure 9.10 represents models of two of the stakeholders – a service provider and an infrastructure provider. Note that we can define each of these agents as a class of agents, which means we can have multiple object instantiations of both (and other) classes of agents.

9.4.2 EXAMPLE 2: ASSESSING NEW TECHNOLOGIES ON FUTURE FLEET AND EMISSIONS

In this problem, there is one key stakeholder, the airline, whose decisions we want to model. Specifically, the focus of this problem is the airline's fleet allocation decisions done in the context of technological, market, economic, and environmental factors; this was graphically represented in Figure 3.12.

To implement this model, we will set up the airline's decision-making as a mixed integer programming problem. Each of the contextual factors will require its own models. Thus, as shown in Figure 3.12, we will have models of aircraft technological advancement, evolution of market demand, operating costs of the airline, etc. Figure 9.11 shows the process by which we set up each of the required models. Once we have all the pieces assembled, we run simulations by setting different scenarios as defined by inputs to each of the models in the implementation.

9.4.3 EXAMPLE 3: AIR TRANSPORTATION NETWORK RESTRUCTURING

In the implementation phase of our final problem, we develop (program) each of the required models that we abstracted out in the previous phase. Refer back to Figure 5.5. This figure shows a generic network growth algorithm. In this problem, the

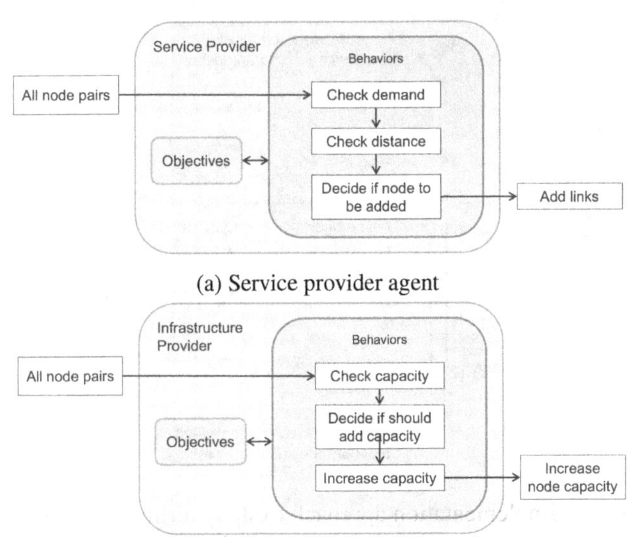

(a) Service provider agent

(b) Infrastructure provider agent

Figure 9.10: Agent models for service provider (at the β-level of the ROPE framework) and infrastructure provider (γ-level).

network of routes will change as a result of addition of new routes or removal of old routes. Thus, the same schematic as that in Figure 5.5 will represent our approach, though "network growth" will be replaced by "network change." The logic for change will be applied to the central block in this figure, also shown as "network growth models" in Figure 9.5.

Specifically, our implementation is shown in Figure 9.12, following the work of Kotegawa and collaborators [111]. We begin with an initial model of the network, which can be obtained from historical data. We feed this model along with definitions of future scenarios to a route forecast algorithm which provides us with traffic volume forecasts on the new route in the network. At the same time, a separate algorithm forecasts traffic at each airport which is also added to the routes. Finally, using all traffic information, the set of new routes and the resulting network is identified, which then becomes the initial network for the next time step of evolution.

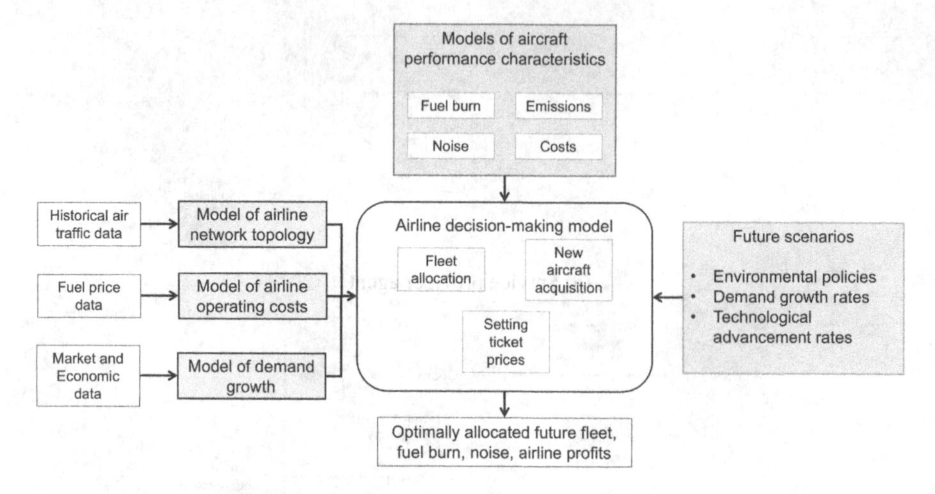

Figure 9.11: Implementation approach to study airline decision making.

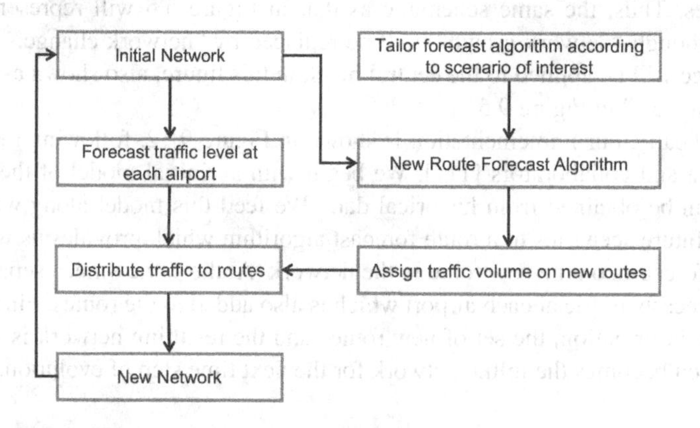

Figure 9.12: Implementation approach to study network restructuring problem.

10 Human Space Exploration System of Systems

In this chapter, we introduce an example of modeling and analysis of the architecture of human space exploration, in particular focusing on Earth-to-Mars missions. As mentioned in Chapters 1 and 2, SoSs pose a challenge because their features often meet with a lack of appropriate approaches and methodologies. In particular, Systems Engineering frequently fails to keep into account aspects of long-term operational life, which also entails a larger search space, with additional uncertainty on stakeholders' needs, policies, and economics.

Human space exploration is an interesting and useful example of SoS for what concerns these aspects. The space community needs new ways to achieve holistic decision-making, due to how the complexity and size of space missions in the near and far future impact risk [51]. In addition, the length of space missions makes them prone to many changes over time, due to the impact of changing long-term policies. Finally, current space endeavors are no longer focusing on single missions or on a sequence of very compartmentalized missions, but on long-term programs, including various integrated efforts, often carried forward by different stakeholders, with many different goals to be achieved.

After introducing the problem of architectural choices in an Earth-to-Mars mission, this chapter shows an application of the DAI process (Chapter 3) to this problem, developed in full detail in [80]. The last section will illustrate different problems in the same area, including evaluation of propulsion systems and atmospheric entry systems, and an example of how to implement considerations on budget and stakeholder preferences in architecting space SoS.

10.1 PROBLEM INTRODUCTION

Systems Engineering in the space field already has many well-established procedures for designing space systems [170]. However, some aspects that arise due to the SoS-like nature of crewed Mars exploration architectures are not well addressed by these procedures. The challenges that this example addresses are:

1. Add top-down approach to Mars exploration SoS architecting.
    a. How do systems interact in the operational domain and contribute to achieving the desired capabilities?
    b. What design principles can improve robustness at the level of a whole, complex series of missions?
2. Evaluate alternative architectures.

a. Which architecture can improve the overall resilience?
b. What is the holistic impact of partial or total failures, accounting for the dependencies?
c. What is the impact of delays and uncertainties on the mission timeline and success?

3. Achieve better, explicit assessment of holistic risk.
   a. Which systems are the most critical to the mission life cycle?
   b. How should partial capabilities evolve over time?
4. Consider that multiple stakeholders are involved.
   a. What is the impact of stakeholder decisions and priorities on the development of the SoS?

Application of the formal DAI approach for SoS (described in Chapter 3) allows the user to obtain useful answers to these questions, keeping into account long-term impact of current decisions, coupling among uncertainties and risks, dependencies between elements, and various constraints. As usual for SoS, due to the size of the problem, the amount of uncertainty, and the openness of some questions, modeling and analysis is not used to provide a full optimization of a problem. Instead, an SoS approach provides support to decision-makers in the early stage of design and architecture, in particular by providing answers to *what-if* questions in specific scenarios, by facilitating risk analysis, and by providing requirements and constraints to the traditional *bottom-up* Systems Engineering approach.

10.2 DEFINITION PHASE

This part of the process is used to identify the constituent parts of the SoS and their hierarchical organization, and to determine the status quo, stakeholders, and disruptors related to the problem. This example addressed the evaluation of robustness and resilience of Mars crewed exploration architecture.

10.2.1 HIERARCHY OF THE MARS EXPLORATION ARCHITECTURES

SoS are usual characterized by multi-level hierarchy among the systems involved. The first step to execute the definition phase for this problem is identifying and developing a hierarchical representation of the Mars exploration architecture SoS. According to the DAI process, each level of abstraction in this hierarchy is indicated with a letter of the Greek alphabet. For the Mars exploration SoS, we can define these levels. Usually the levels are defined starting from the α level (that is, the lowest level of abstraction), whose composition is chosen based on the smallest entities that it is necessary to model in order to answer the proposed questions. However, for ease of understanding, they are here ordered from the highest to the lowest level of abstraction:

- **γ level**
  This is the highest — and least detailed — level of abstraction for this application. This level, which stands above the largest monolithic systems

of the Mars exploration architecture, consists of high-level functions and capabilities, to which sets of systems will be allocated. This level also includes policies, economic factors and stakeholder decisions relative to the functions and capabilities. Due to the low amount of detail, values at this level are useful to compare the performance of different architectures, that might share common objectives but a substantially different decomposition at lower levels.

- **β level**
  In the proposed hierarchical taxonomy for this application, this is the intermediate level of abstraction. This level contains large complex monolithic systems, for example the Orion Spacecraft, the Space Launch System (SLS), a whole control center, a set of communication satellites. Smaller scale functions and capabilities, related to individual large systems, may also be listed in this level. Considerations about policies, economics, and stakeholders related specifically to these monolithic systems are also part of this level.
- **α level**
  This is the lowest — and most detailed — level of abstraction for this application. The systems in the β level are decomposed into on-board systems and facilities subsystems. Examples of entities at this level are communication antennas, solar arrays, the structural components of a rocket, spacecraft payloads, etc. Consideration about subsystems policies, economics, and stakeholders are also part of this level.

If the questions that the user is trying to answer require more detailed models, this hierarchical decomposition can be expanded into lower levels. When these entities are treated in the abstraction phase, their dependencies are used to organize entities in dependency networks at various levels of abstraction.

Since the definition phase is meant to increase the understanding of the underlying problem to be addressed, in this phase the user defines the taxonomy of the systems and stakeholders involved, and the technological needs. The user also identifies the operational context, the status quo, the domain of application, time scales, desired goals for the future, and improved architectures. Finally, expected barriers to the preferred behavior are identified in this phase.

As described in Chapter 3, one important product of this phase is the ROPE table, which contains the Resources, Operations, Policies, and Economics (ROPE) at various hierarchical levels for the problem under study. Table 10.1 is the ROPE table for Mars exploration architecture SoS.

It is important to underline that the same ROPE table can be used to drive different SoS problems, usually involving a limited number of hierarchical abstraction levels and of ROPE aspects. In this study, choices regarding different architectures, the technologies involved, contractors and partners, and the goals will be reflected into the development time and above all in the uncertainties both in the developmental and operational domain. In the case of Mars exploration, the user also needs to take into account the launch windows. The cheapest direct trajectories, using Hohmann

Table 10.1
ROPE Table for the Mars Exploration Architecture SoS

Levels	Resources	Operations	Policies	Economics
γ	High-level functions, concept architectures	High-level disruptions and drivers, goals, contractors, partners	Launch windows, timelines, partnership with commercial enterprises	Budget allocation, drivers for partners, cost of operations
β	Control centers, ground support, comm networks, In-Situ Resource Utilization, large complex systems, system-level capabilities	Systems disruptions, systems development, contractors, partners, orbit and trajectories	Location of facilities, architectural and technological choices (e.g., assembly in Moon orbit, Solar Electric Propulsion), systems interactions	Budget allocation, cost of development and operation
α	On-board systems and subsystems (e.g., solar panels, controllers), low-level capabilities	Subsystems goal and possible disruptions	Architectural and technological choices, subsystems interactions	Low level cost and budget allocation.

transfers, result in launch windows separated from each other at intervals of slightly more than two years (the synodic periodic between Earth and Mars is about 780 days). Table 10.2 lists 12 launch windows between 2018 and 2041, calculated for a trip of 6 months.

Since the SoS problem addressed in this example is a space exploration concept with a specific destination, literature review in the definition phase includes a survey of official concepts. Starting from 1948, when Wernher Von Braun proposed *Das Marsprojekt*, translated and published in English in 1953 [166], about 70-80 concepts for Mars exploration have been proposed between the twentieth and the twenty-first centuries. Based on large-scale features, the concepts can be sorted in categories and subcategories, which will be useful in the abstraction and implementation phases to drive the comparison between alternative architectures. The following list uses two categories and five total subcategories.

Table 10.2
Twelve Launch Windows for Direct Launch to Mars between 2018 and 2041, and Associated Missions Launched or in Development

Launch window number	Date	Scheduled launches
1	April 2018	InSight (NASA, JPL) Red Dragon (SpaceX, cancelled)
2	July 2020	Emirates Mars Mission - Hope (UAE) Tianwen-1 (CNSA) Mars 2020 (NASA, JPL)
3	September 2022	Exomars (ESA, Roscosmos)
4	November 2024	Martian Moons Exploration (JAXA) Escapade (NASA) Mangalyaan 2 (ISRO)
5	January 2027	
6	February 2029	
7	April 2031	
8	June 2033	
9	July 2035	
10	September 2037	
11	October 2039	
12	December 2041	

NASA: National Air and Space Administration, JPL: Jet Propulsion Laboratory, UAE: United Arab Emirates, CNSA: China National Space Agency, ESA: European Space Agency, JAXA: Japanese Aerospace Exploration Agency, ISRO: Indian Space Research Organization

1. Single mission from Earth to Mars (crew, cargo, and possible return vehicle sent in a single launch to Mars). The mission can be repeated, but each launch will carry all typology of payloads
   a. Assembly in LEO or Moon Orbit (e.g., Project Empire)
   b. Heavy-lift launch from Earth (e.g., Martian Piloted Complex, shown in Figure 10.1, SpaceX Mars Colonial Transporter)
2. Separate launches to Mars (multiple missions to bring separate cargo, vehicles, and crew to Mars)
   a. Direct launches to Mars, with return (e.g., Mars Direct)
   b. Direct launches to Mars, without return (e.g., Mars-to-stay, Mars One)
   c. Separate launches with intermediate missions (e.g., NASA Constellation, Space Exploration Initiative, Mars Cyclers, NASA Design Reference Mission 5.0, Figure 10.2, Journey to Mars)

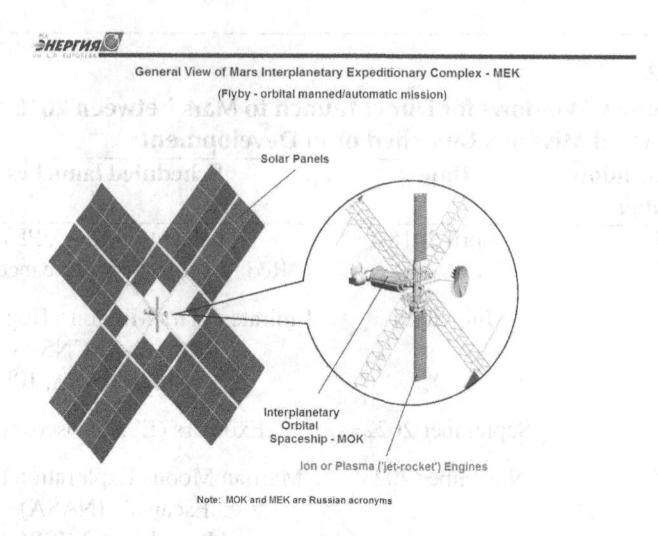

Figure 10.1: Martian Piloted Complex, by RKK Energia. Interplanetary spacecraft, with compartments for living, working, exercising, farming.

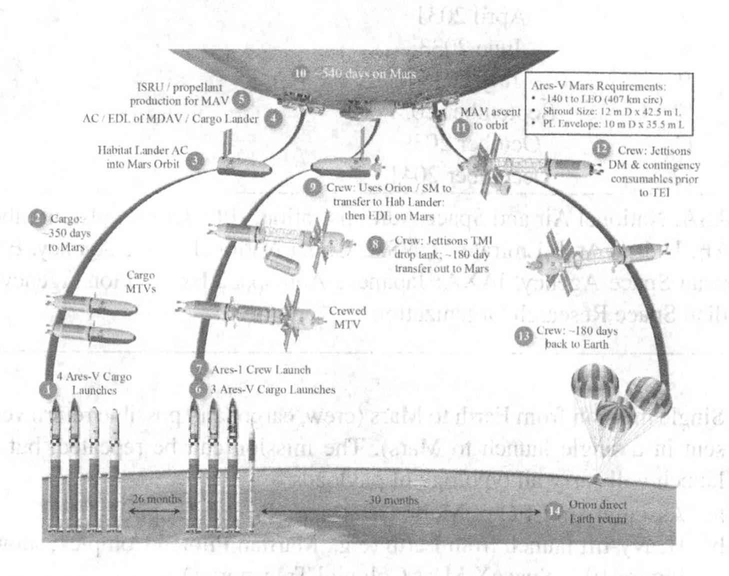

Figure 10.2: Mission profile of NASA Design Reference Mission Architecture 5.0.

Each of the concepts can be implemented with different architectures, based on the systems allocated to functions and capabilities, technological and economic constraints, and possible sequential missions or steps required to finally achieve a mission to Mars. By architecture we mean the organization of systems and capabilities in dependency networks, accounting for stakeholder decisions, policies, economics, technology requirements, trajectories, and objectives (that is, items belonging to the three classes of variables described in Section 3.4: Composition, Configuration, and Control).

This example will focus on architectures in only two different categories: individual missions carrying both crew and cargo, each one executed with a single heavy launch from Earth (category 1b) and a mission composed of separate launches with intermediate objectives (category 2c). The architecture in category 1b (architecture A) will have multiple stand-alone missions, each performed with a heavy vehicle which will carry crew, cargo, equipment, and re-entry vehicle directly from Earth to Mars. The architecture in category 2c (architecture B) will involve separate cargo trips to Mars or the vicinity of Mars using Solar Electric Propulsion, habitat assembly in Moon orbit, construction of bases on the Moon, crew launched to Mars with Space Launch System (SLS) rocket and Orion spacecraft (chemical propulsion), after transitioning into Moon orbit.

Each category of architectures will require specific technologies and capabilities. In this phase, we can also identify technologies and considerations pertaining to each of the five categories, so as to assess the context and the status quo, and guide the abstraction and implementation phases.

- Common technologies across the categories
  - Control centers (locations, stakeholders, authorities)
  - Ground support (development, test, construction, integration, operation)
  - Communication networks
  - In-Situ Resource Utilization (ISRU)
  - Physical and psychological impact of mission (including radiation shielding, etc.)
- Category 1a (Single missions to Mars – Assembly in LEO or Moon Orbit)
  - Rendezvous, docking, and extended assembly in space
  - Medium to heavy lifters for assembly missions
  - Windows for timely launches of components and assembly missions
  - Heavy lifter
  - Chemical / hybrid / nuclear propulsion (with crew)
  - Fast trajectory towards Mars
  - Piloted Entry, Descent and Landing (EDL) of super-heavy vehicle
- Category 1b (Single missions to Mars – heavy lift launch from Earth)
  - Super-heavy lifter
  - Chemical / hybrid / nuclear propulsion (with crew)
  - Fast trajectory towards Mars
  - Piloted EDL of super-heavy vehicle
- Category 2a (Separate cargo and crew launch, direct launches from Earth to Mars, with return to Earth)
  - Medium to heavy lifters
  - Chemical / hybrid propulsion for crew vehicle
  - Solar Electric Propulsion / Nuclear Electric Propulsion / hybrid for cargo vehicles

- Windows for timely launches of the components
- Fast trajectory towards Mars for crew vehicle
- Cheap and/or fast trajectories towards Mars for cargo and vehicles
- Multiple automated and accurate EDLs
- EDL of re-entry vehicle
- Piloted and accurate EDL for the crew
- In-situ production of water and fuel

- Category 2b (Separate cargo and crew launch, direct launches from Earth to Mars, without return to Earth)
  - Medium to heavy lifters
  - Chemical / hybrid propulsion for crew vehicle
  - Solar Electric Propulsion / Nuclear Electric Propulsion / hybrid for cargo vehicles
  - Windows for timely launches of the components
  - Fast trajectory towards Mars for crew vehicle
  - Cheap and/or fast trajectories towards Mars for cargo and vehicles
  - Multiple automated and accurate EDLs
  - Piloted and accurate EDL for the crew
  - Extended structures and In-Situ Resource Utilization for colonization
- Category 2c (Separate cargo and crew launches, intermediate missions with assembly in space or on other bodies, with return to Earth)
  - Assembly in space or on other bodies
  - Sequential mission launch windows and trajectories
  - Medium to heavy lifters
  - Chemical / hybrid propulsion for crew vehicle
  - Solar Electric Propulsion / Nuclear Electric Propulsion / hybrid for cargo vehicles
  - Windows for timely launches of the components
  - Development of cycler vehicles, and high-velocity rendezvous
  - Fast trajectory towards Mars for crew vehicle from different origin (Earth, Moon, Near Earth Objects, Cyclers)
  - Cheap and/or fast trajectories towards Mars for cargo and vehicles
  - Multiple automated and accurate EDLs
  - EDL for re-entry vehicle
  - Piloted and accurate EDL for crew
  - In-situ production of water and fuel
  - Windows for re-entry
  - Extended structures and In-Situ Resource Utilization for colonization

10.3 ABSTRACTION PHASE

Through iteration with the definition and the implementation phase, abstraction was mainly focused on the generation of appropriate dependency networks of the systems, technologies, and capabilities identified during the definition phase. These dependencies are both in the operational and developmental domain, and have been generated at various levels of abstractions and for different architectures. Most of this work, driven by the information gathered in the previous phase, is based on literature review and interactions with Subject Matter Experts (SMEs).

10.3.1 γ LEVEL

In this application, the nodes of the networks of operational and developmental dependencies networks at this level are capabilities and overarching goals of the architectures for Mars exploration. The outcomes that the user wishes to analyze and methodologies like the Design Structure Matrices (DSM) [32] can be combined with the elements identified in the definition phase, to drive the decision of nodes to include at the various levels. Figure 10.3 shows the developmental and operational dependencies between capabilities at the γ level in architecture A. Figure 10.4 shows the developmental and operational dependencies between the capabilities at the γ level in architecture B.

Architecture B, which belongs to the category of concepts based on a series of intermediate steps to reach Mars, has more capabilities that need to be developed. However, this is not enough to decide that architecture A is preferable, since the capabilities in architecture A may require more effort in terms of time and cost, and may be less robust and less flexible.

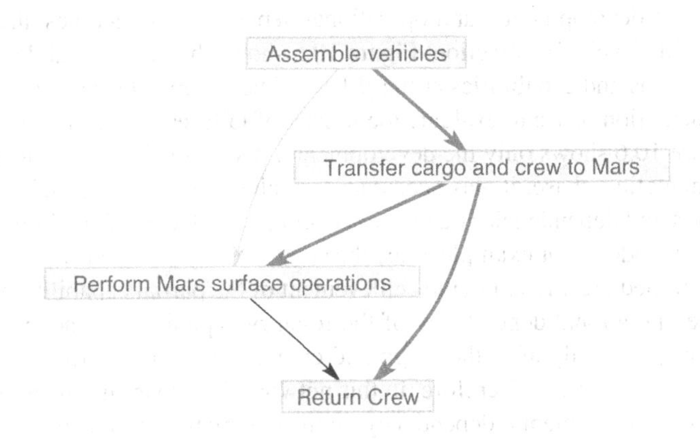

Figure 10.3: Developmental and operational dependencies at the γ level in architecture A. Thin black edges: developmental dependencies. Thin light edges: operational dependencies. Thick edges: both kinds of dependencies.

10.3.2 β LEVEL

Given the proposed questions on comparing architectures, in this application, the networks of operational and developmental dependencies at this level are the core

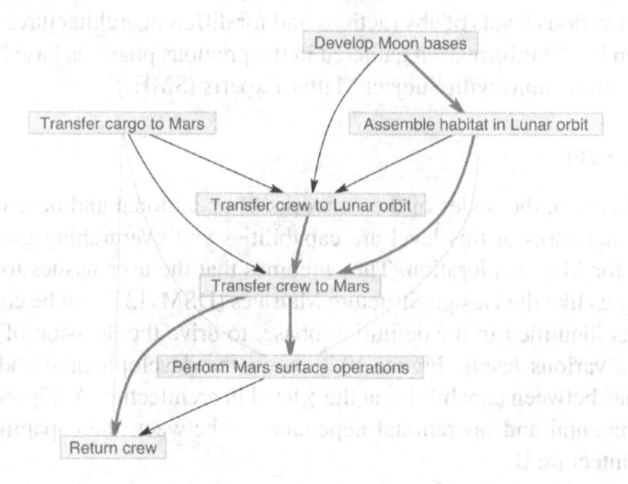

Figure 10.4: Developmental and operational dependencies at the γ level in architecture B. Thin black edges: developmental dependencies. Thin light edges: operational dependencies. Thick edges: both kinds of dependencies.

of the abstraction and implementation phases. For both architectures under study, the network of developmental and operational dependencies becomes already quite complex at this level of abstraction. Figure 10.5 shows the operational dependencies between systems and capabilities at the β level. These dependencies will be used in the implementation phase to evaluate the impact of failures on the capabilities of interest. Figure 10.6 shows only the developmental dependencies. After iteration with the implementation phase, it was necessary to include both dependencies due to required inputs and dependencies due to sequencing (which caused the introduction of some dummy nodes). For example, once the crew re-entry vehicle is fully developed, it does not immediately result in the achievement of the partial capability of re-entry, because the operational dependency of the re-entry capability on the crew re-entry vehicle takes place only after the cargo and crew have been deployed on Mars and have executed their tasks. Therefore, in this network the capability of cargo delivery to Mars has a developmental dependency on the completion of the crew re-entry vehicle, and the capability of re-entry had a developmental dependency on the cargo delivery to Mars.

Figures 10.7 and 10.8 show analogous dependencies for architecture B.

10.3.3 α LEVEL

For this example, some of the systems at the β level have been decomposed into subsystems and development phases, to show an example of the operational and developmental dependencies at the α level of abstraction. Due to the scope of this case study, no analysis has been performed at this level in the implementation phase. However, to give an example of products of the abstraction phase for this

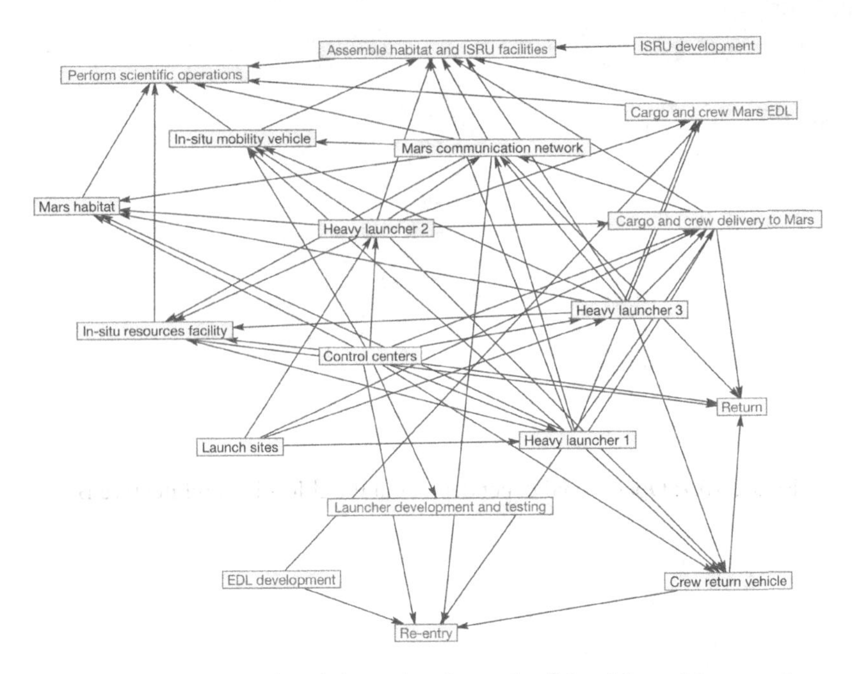

Figure 10.5: Operational dependencies at the β level in architecture A.

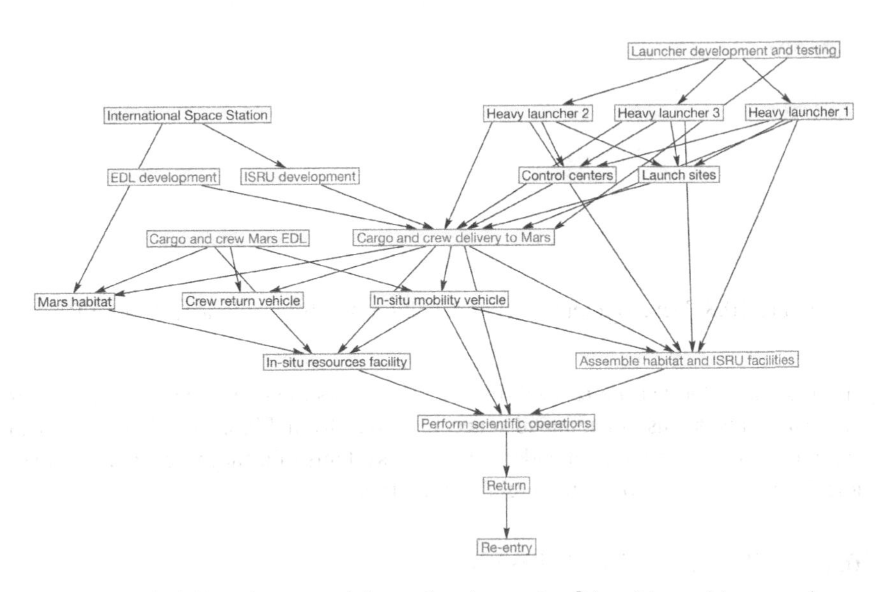

Figure 10.6: Developmental dependencies at the β level in architecture A.

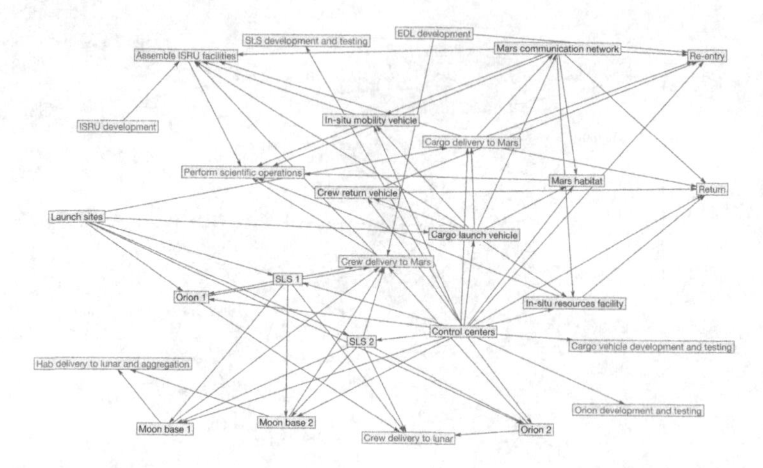

Figure 10.7: Operational dependencies at the β level in architecture B.

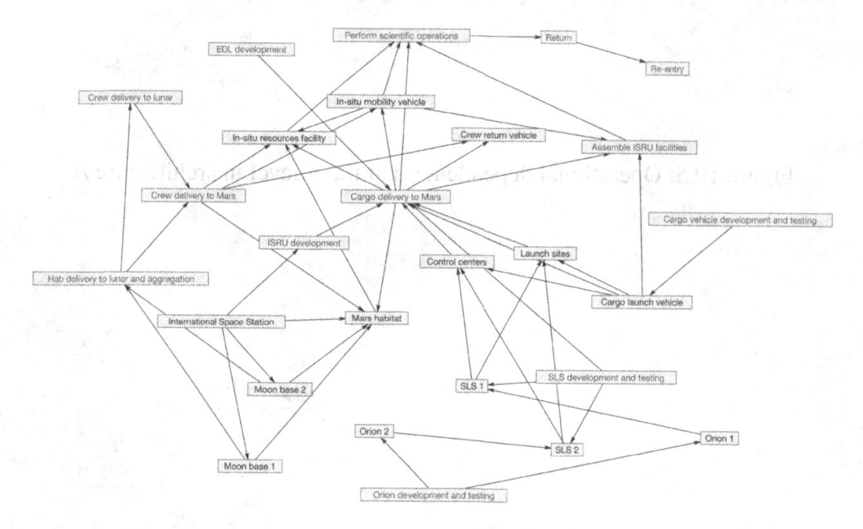

Figure 10.8: Developmental dependencies at the β level in architecture B.

particular case study at the α level, Figure 10.9 shows the operational dependencies among the subsystems of a manned spacecraft, like Orion. Figure 10.10 illustrates an example of developmental dependencies of the systems of a cargo transportation vehicle developed in part by commercial contractors.

10.4 IMPLEMENTATION PHASE

For this problem, implementation addressed the β level dependencies. Nodes at the γ level in developmental dependencies networks have not been characterized with their own development time. Their sequencing is implemented based on the development time and dependencies of the corresponding β level structure. Likewise,

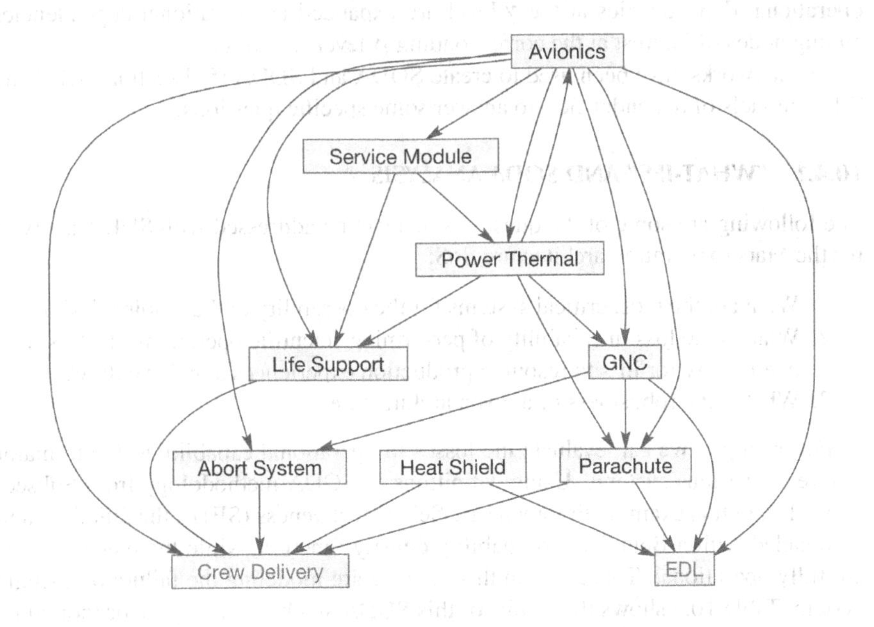

Figure 10.9: Operational dependencies at the α level of a manned spacecraft.

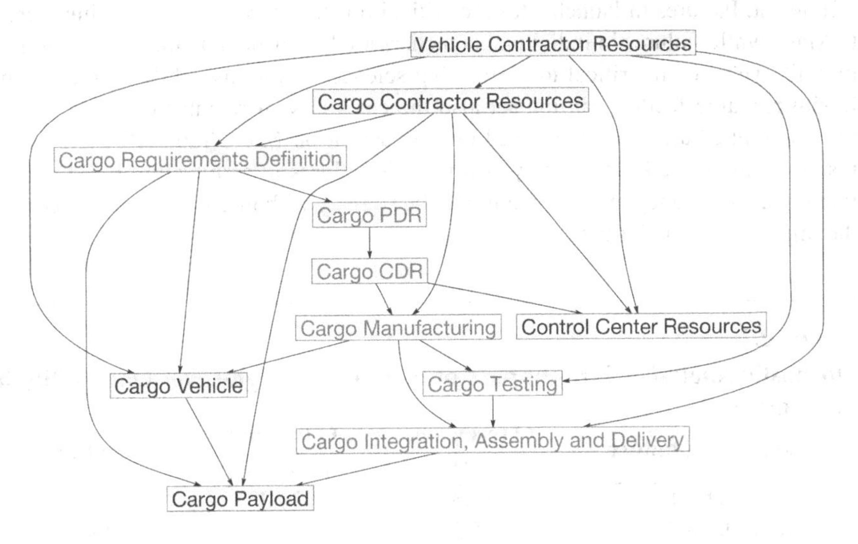

Figure 10.10: Developmental dependencies at the α level of a cargo transportation vehicle, developed in part by a commercial contractor. PDR: Preliminary Design Review. CDR: Critical Design Review.

operational dependencies at the γ level are expanded as operational dependencies among nodes of interest at the corresponding β level structure.

The networks have been used to create SODA and SDDA (Subsections 7.1.2 and 7.1.4) models of dependencies, to answer some specific questions.

10.4.1 "WHAT-IFS" AND SODA ANALYSIS

The following are some of the questions that can be addressed with SODA analysis for the Mars exploration architecture SoS:

1. What are the most critical systems for the operability of the whole mission?
2. What is the loss in capability of performing scientific operations of Mars if one facility for in-situ resource production experiences a major failure?
3. What is the robustness of a given architecture?

For example, we can evaluate the losses in operational capabilities due to major failures in various systems. Using definitions of SODA methodology from Subsection 7.1.2, in this example the stochastic Self-Effectiveness (SE) of the failed system is modeled with a Beta(2,11) probability density function, while the other systems are fully operational. Therefore, in this case we are modeling the failure of a single system. Table 10.3 shows the results of this SODA stochastic analysis for individual system failures in architecture A. The table shows the resulting operability (ranging from 0 to 100, with 0 being a total loss and 100 perfect functionality) of capabilities of interest. Failures in launch sites are critical for the operability of carrying a crew to Mars, while failures in individual launchers are less critical. Failures in the in-situ mobility vehicle are critical to performing science operations, while failures in the in-situ resource facility can impact also the capability of re-entry on Earth, which in this architecture is in part based on production on Mars. Table 10.4 shows the results of the same kind of analysis for single failures in architecture B, where the increased redundancy of the systems and the strategy with intermediate steps reduces the impact of single failures.

Table 10.3
Stochastic Analysis of the Impact of Failures on High-Level Operability in Architecture A

System with failure	O (crew on Mars)	O (science operations)	O (return)
Launch sites	14.8	74.3	37.3
Heavy launcher 1	95.5	95.4	96.1
In-situ resource facility	100	96.0	91.4
In-situ mobility vehicle	100	89.2	100
Crew return vehicle	100	100	83.7

While single failures have more impact on the outcome of architecture A, the presence of a larger number of necessary systems in architecture B causes a slightly

Table 10.4
Stochastic Analysis of the Impact of Failures on High-Level Operability in Architecture B

System with failure	O (crew on Mars)	O (science operations)	O (return)
Launch sites	79.7	71.9	51.8
SLS 1	93.2	99.9	100
Orion 1	94.3	99.2	100
Moon base 1	97.1	99.6	100
In-site resource facility	100	96.9	93.4
In-situ mobility vehicle	100	91.2	100
Crew return vehicle	100	100	97.8

lower amount of robustness when including multiple failures. The values, computed according to Equation 7.35, are as follows.

$$Rob_A = 0.39$$

$$Rob_B = 0.37$$

10.4.2 "WHAT-IFS" AND SDDA ANALYSIS

The following are some of the questions that can be addressed with SDDA analysis for the Mars exploration architecture SoS:

1. What is the delay (including possible missed launch windows) and impact on capabilities if a launcher experiences a delay during development?
2. What are the most critical systems in terms of development delays?
3. What if the initial assessment of development time is wrong? Which development policy is the most flexible and the safest?

The SDDA parameters of this implementation reflect the complexity and the Technology Readiness Level (TRL) of the systems involved. Launch windows occur at a distance of 110 weeks from one another; therefore, uncertainties and delays can have a higher impact, because the resulting architecture may miss a launch window.

Figure 10.11 shows the baseline schedule resulting from the developmental dependencies for architecture A, and Figure 10.12 shows the baseline schedule resulting from the developmental dependencies for architecture B. Due to launch windows, dependencies and times of development, both architectures show crew return at the same time in their baseline schedule (about 12 years). Architecture A appears to have more flexibility and spare time before the first launch, so a possible capability to absorb more delays. However, it is also the architecture whose systems are expected to have higher impact if delays arise, due to their size and to the absence of intermediate steps during development.

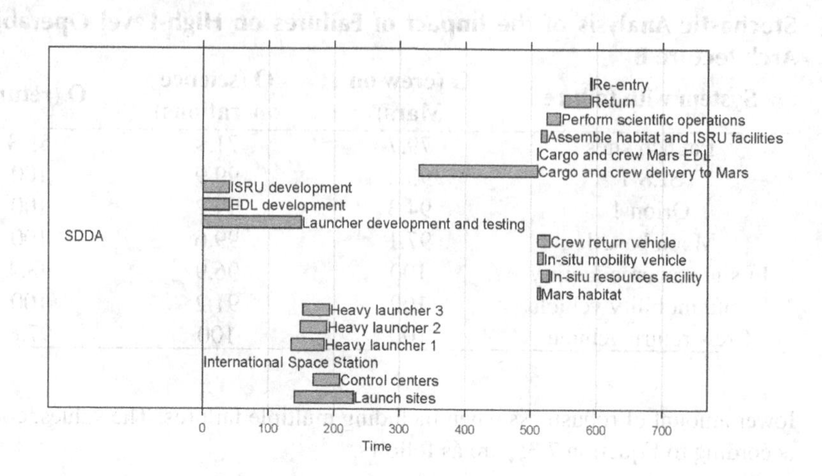

Figure 10.11: Baseline schedule of the β level of architecture A for Mars exploration.

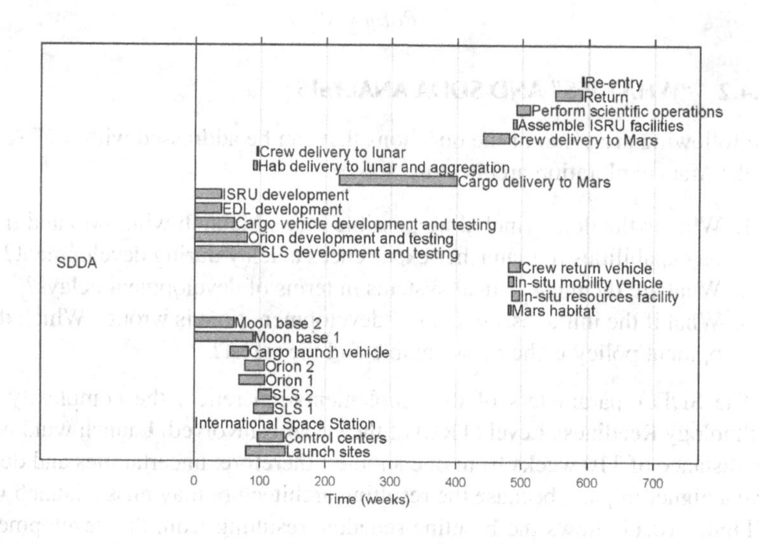

Figure 10.12: Baseline schedule of the β level of architecture B for Mars exploration.

Deterministic SDDA analysis assesses the criticality of the systems in the developmental domain. Table 10.5 shows the effect of delays in individual systems or development stages on the overall schedule in architecture A (according to definitions from Section 7.1.4, the punctuality of delayed systems is equal to 10). Due to the discrete launch windows, large groups of systems cause the same impact on the final schedule in this deterministic case. The development and testing of new, heavy launchers are more critical than the final phase of building the launchers. Delays in assembling the habitat and ISRU facilities on Mars have delays on the execution of scientific research, but they are still absorbed before the launch window for return.

Table 10.6 shows results of the same kind of analysis in architecture B. Due to the lower amount of spare time before the first launch in this architecture, the architecture is expected to have less overall robustness. However, since the development of the various technologies can proceed in parallel, and the uncertainty has less impact than in the case of heavy launchers, the architecture is robust for all delays in single systems or developmental stages.

Table 10.5

Deterministic Analysis of the Impact of Delays on the Overall Schedule in Architecture A (weeks)

System with delay	$t_C^{crewonMars}$	$t_C^{scienceoperations}$	t_C^{return}
None	510	544.4	592
Launch sites	510	544.4	592
Heavy launcher 1	510	544.4	592
Heavy launcher 3	510	544.4	592
Crew return vehicle	510	544.4	592
Launcher development and testing	620	654.4	702
ISRU development	510	544.4	592
Assemble habitat and ISRU facilities	510	549.3	592

Table 10.6

Deterministic Analysis of the Impact of Delays on the Overall Schedule in Architecture B (weeks)

System with delay	$t_C^{crewonMars}$	$t_C^{scienceoperations}$	t_C^{return}
None	480	511.6	592
Launch sites	480	511.6	592
SLS 1	480	511.6	592
Orion 1	480	511.6	592
SLS development and testing	480	511.6	592
ISRU development	480	511.6	592
Hab delivery to lunar and integration	480	511.6	592
Assemble ISRU facilities	480	518.8	592

This deterministic analysis, however, puts all the systems at the same level of disrupted punctuality, which is usually not the case. To quantify the probability that each architecture will follow the expected schedule, it is more realistic to use stochastic analysis, with levels of uncertainty based on the complexity and Technology Readiness Level (TRL) of each system and stage. For this case, 10000 runs of stochastic SDDA analysis have been used. In architecture A, due to the complexity of the systems involved, and the absence of intermediate steps, the uncertainty is large and causes delays beyond the first available launch window in many cases. In architecture B, the uncertainty of the schedule of individual systems and stages is more variable, resulting in a lower number of instances with delays, but some of the specific delays can miss two or even three launch windows.

Tables 10.7, 10.8, and 10.9 show percentage of instances that completed tasks within a given time threshold in the two architectures.

Table 10.7
Percentage of Instances Completing Delivery of Crew on Mars within Given Time Thresholds

	Crew arrival on Mars					
	$t_C < 500$	$500 \leq t_C < 550$	$550 \leq t_C < 600$	$600 \leq t_C < 650$	$650 \leq t_C < 700$	$700 \leq t_C$
Arch. A	0 %	7.2 %	2.7 %	36.5 %	53.6 %	0 %
Arch. B	81.1 %	0.01 %	11.2 %	7.4 %	0 %	0.3 %

Table 10.8
Percentage of Instances Completing Scientific Portion of the Mission on Mars within Given Time Thresholds

	Perform scientific operations					
	$t_C < 550$	$550 \leq t_C < 600$	$600 \leq t_C < 650$	$650 \leq t_C < 700$	$700 \leq t_C < 750$	$750 \leq t_C$
Arch. A	0 %	8.2 %	1.7 %	51.5 %	38.6 %	0 %
Arch. B	81.2 %	0 %	18.5 %	0.07 %	0.07 %	0.2 %

10.4.3 "WHAT-IFS" AND COMBINED SODA/SDDA ANALYSIS

The following are some of the questions that can be addressed with combined SODA and SDDA analysis for the Mars exploration architecture SoS:

1. How do capabilities change over time?

Table 10.9
Percentage of Instances Completing the Return from Mars within Given Time Thresholds

	Return					
	$t_C < 600$	$600 \leq t_C < 650$	$650 \leq t_C < 700$	$700 \leq t_C < 750$	$750 \leq t_C < 800$	$800 \leq t_C$
Arch. A	0 %	0 %	0 %	9.9 %	0 %	90.1 %
Arch. B	31.4 %	49.8 %	0 %	18.6 %	0 %	0.2 %

2. How to decide if, when, and how to update the SoS (choice of involved stakeholders, cost, improvement, and risk)?

This is an example of combined use of SODA and SDDA in the baseline case, where no failures or delays occur. Figure 10.13 shows the evolution of the operability of required capabilities over time in architecture A, and Figure 10.14 shows the evolution of the operability of required capabilities over time in architecture B. Both architectures show two major steps, corresponding to the completion of the cargo transportation vehicles with the first launches, and to the arrival of crew on Mars and assemblage of the required facilities. However, architecture A reaches the first step after development and completion of the heavy launchers, which transport both cargo and crew together in subsequent launches. In architecture B, various technologies are developed in parallel, reaching some plateau when operations are moved from the Earth to the Moon. The cargo is sent separately, resulting in partial achievement of capabilities on Mars. When the crew arrives on Mars, the second major step occurs. This behavior confirms the previous analysis, showing that architecture B requires more technological improvements, thus slightly decreasing the robustness, but it also shows intermediate capabilities which improve the flexibility of the architecture and react better to delays. This simple analysis can be expanded with the introduction of deterministic or stochastic failures and delays, stakeholder policies (modeled similarly to how launch windows are modeled, i.e., causing not just delays but entire shifts in the development of some system), and considerations about cost. Finally, the results of SODA and SDDA high- and medium-level analysis provide constraints and requirements for detailed systems engineering at the lower level.

10.5 BEYOND THE INITIAL SOS

As mentioned in the discussion on the definition phase, a complete ROPE table can be used to address different SoS problems. After providing an example of comparison of categories of architectures at a high level, the next two parts illustrate different studies executed on the same ROPE table, with different abstraction and implementation. These studies address different aspects of space exploration SoS problems.

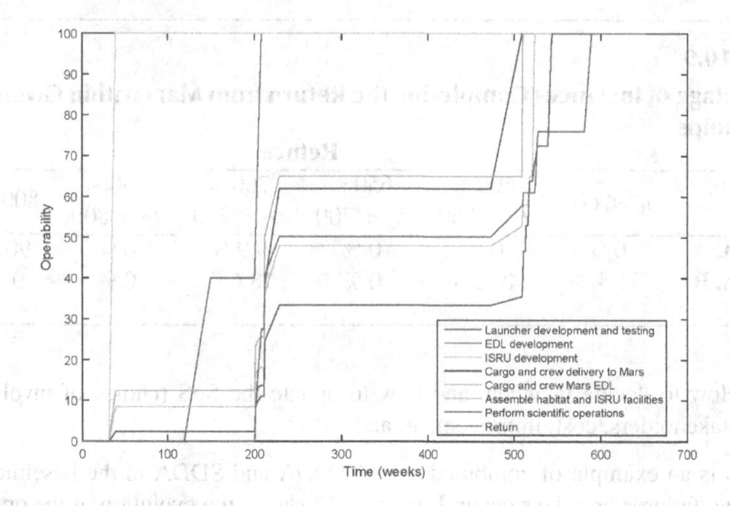

Figure 10.13: Combined SODA and SDDA analysis to quantify the evolution of capabilities over time in architecture A.

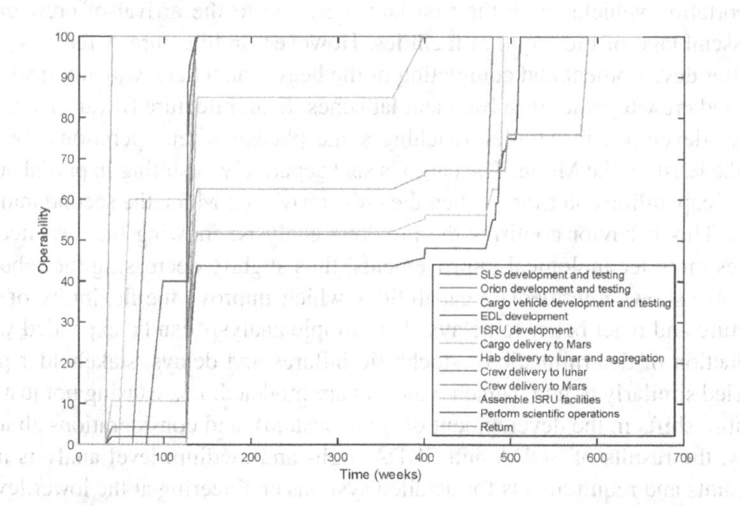

Figure 10.14: Combined SODA and SDDA analysis to quantify the evolution of capabilities over time in architecture B.

10.5.1 LOWER LEVEL OF ABSTRACTION: PROPULSION SYSTEMS AND LUNAR GATEWAY HABITAT SUBSYSTEMS

In this case, the research questions addressed components of the SoS down to the α level. Therefore, in the abstraction phase, further models were developed to expand the network of dependencies between components.

10.5.1.1 Propulsion Systems

The purpose of this application is to compare criticalities, cost, and schedule of three types of in-space propulsion systems: chemical, Solar Electric Propulsion (SEP), and Nuclear Thermal Rockets (NTR). Figure 10.15 shows an example of the dependencies between subsystems in a chemical propulsion system.

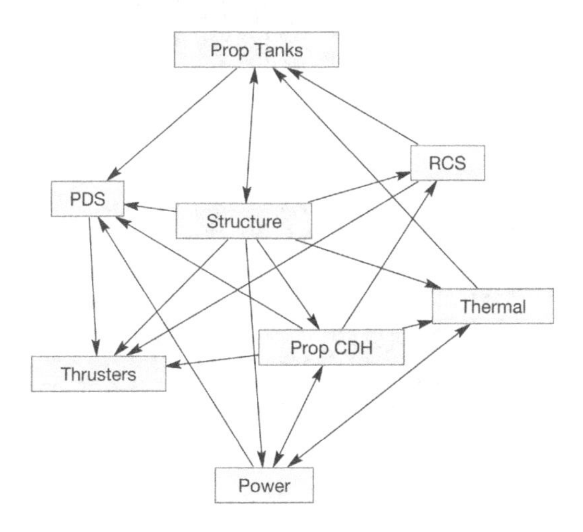

Figure 10.15: α level dependency network for a chemical propulsion system.

Combined analysis with RPO and SODA (Subsections 7.1.1 and 7.1.2) identified optimal portfolios of potential subsystems for each type of propulsion systems and assessed the operational criticalities. Figure 10.16, produced by implementation with SODA analysis for optimal systems in each type of propulsion, shows the expected operability of each subsystem, based on historical data and literature review on functionality of subsystems, and on the operational dependencies of the subsystems. In all propulsion systems, the thermal control subsystem has the lowest expected operability (that is, it is more prone to partial or total disruptions). The resulting expected operability of the subsystems of interest is consistent with historical data.

Further SODA analysis assesses the impact of disruptions in individual subsystems on the operability of the whole propulsion system. Results are shown in Figure 10.17. These results are very different for each of the propulsion systems: in chemical propulsion, the reaction control systems and structure are the most critical subsystems; in NTR, thermal control and power distribution are the most critical subsystems; finally, in SEP, structure and power are the most critical.

These results have then been combined according to the dependency networks developed during the abstraction phase, to provide analysis at a higher level of abstraction. Finally, developmental dependencies and analysis of the impact of delays in subsystems completed the comparison of full architectures characterized

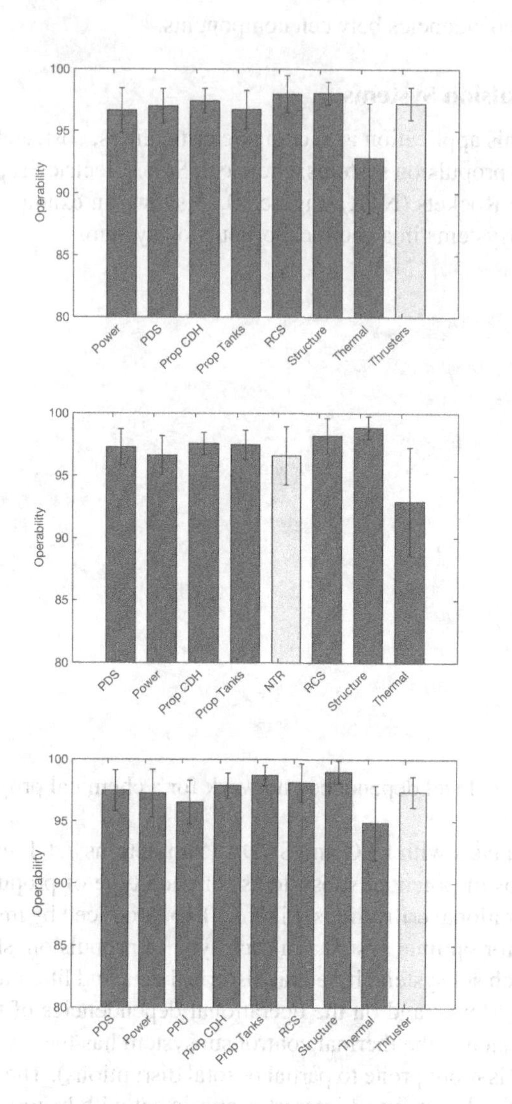

Figure 10.16: Expected Value of operability of subsystems in propulsion systems. Top: chemical. Center: NTR. Bottom: SEP. Light-colored bar is the end node. Error bars indicate 1σ standard deviation.

by different propulsion systems. Table 10.10 shows comparison of optimal architectures that use chemical-based and NTR-based in-space propulsion system. The architecture with NTR is cheaper than the chemical-based, but it requires longer time for first crewed mission and for full mission completion. The architecture with chemical-based propulsion is slightly more robust. In the operational domain,

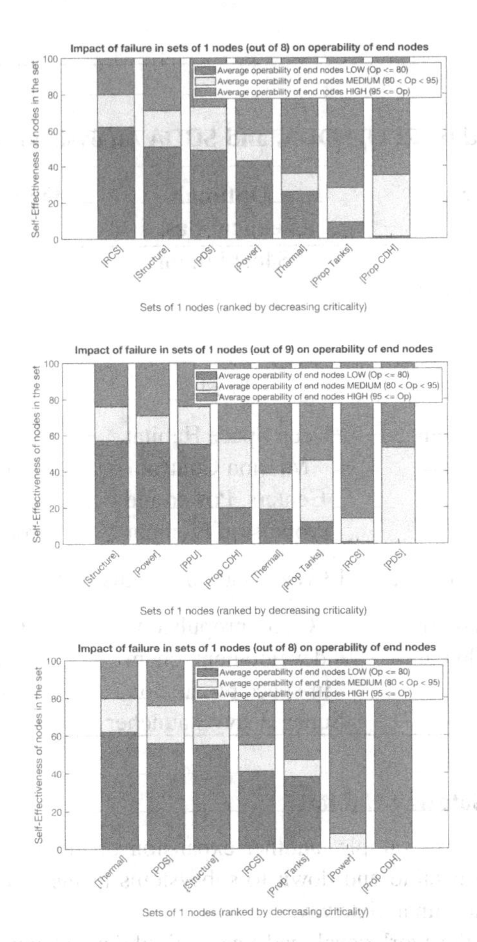

Figure 10.17: Impact of disruptions in the propulsion subsystems. Shades indicate the nominal, sub-nominal, and critical status of the subsystems of interest when the system indicated at the bottom of the bar experiences increasing disruptions, with values of Self-Effectiveness indicated on the vertical axis. Top: chemical. Center: NTR. Bottom: SEP.

critical systems are similar for the two architectures, while in the developmental domain the development and implementation of NTR propulsion has the highest impact on potential delays, while the architecture with chemical-based propulsion is affected by potential delays in the development of a super heavy launch system and of Orion spacecraft.

Full results of this analysis can be found in [87].

Table 10.10
Metrics Generated by RPO, SODA, and SDDA Analysis for Alternative Optimal Architectures

Architecture	Optimal chemical-based	Optimal NTR-based
Cost	$46.4 billion	$40.9 billion
Earliest first crewed mission	10 years	12 years
Number of launches	14	10
E(Op) of mission	98.24	98.11
Most critical system (operational domain)	Deep Space Habitat, Mission Control Centers, Power and Propulsion Element	Deep Space Habitat, Mission Control Centers, In-Space Propulsion System
Expected completion time	13 years and 10 months	18 years and 9 months
Most critical systems (developmental domain)	Chem propulsion development, chem in-space propulsion, Super Heavy Launcher	NTR development, NTR in-space propulsion

10.5.1.2 Lunar Gateway Habitat

Similar to the previous example, another expansion of the abstraction undertaken from the same ROPE table and down to subsystems at the α level addressed the crewed habitat of the Lunar Gateway.

For this purpose, this work developed a network of dependencies between systems of a space habitation module, at a level corresponding to the α level in the ROPE table (Table 10.1). This network is shown in Figure 10.18. However, the table used previously to answer questions about the decision-making process at higher levels of abstraction can be further expanded to lower levels, if necessary to address the questions of interest.

Figure 10.19 show the expansion of the systems of the space habitation module in 27 subsystems. This study modeled and implemented the interdependencies between subsystems to evaluate the impact of failures on the operational capabilities of the various parts of the Gateway habitat.

Extended simulation provided information on the impact of disruptions in various individual subsystems or sets of subsystems on the operational capabilities of each other part of the habitat. For example, Figure 10.20 shows the impact of individual failures in each of the subsystems on the atmosphere management, which is a subsystem of the Environmental Control and Life Support System (ECLSS). Results show that the fire safety subsystem has the largest impact on the correct operation of the

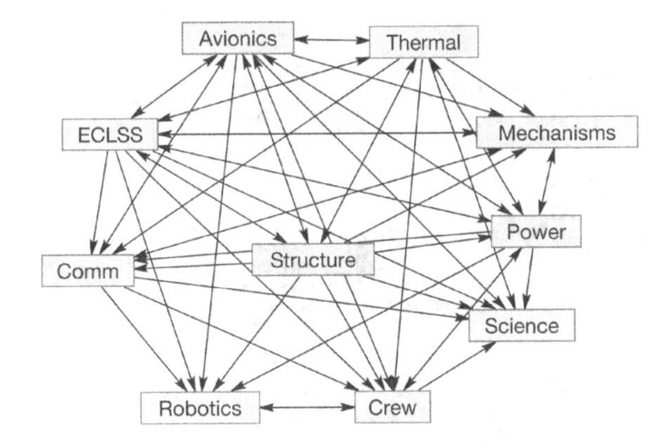

Figure 10.18: Interdependencies between systems of a space habitation module.

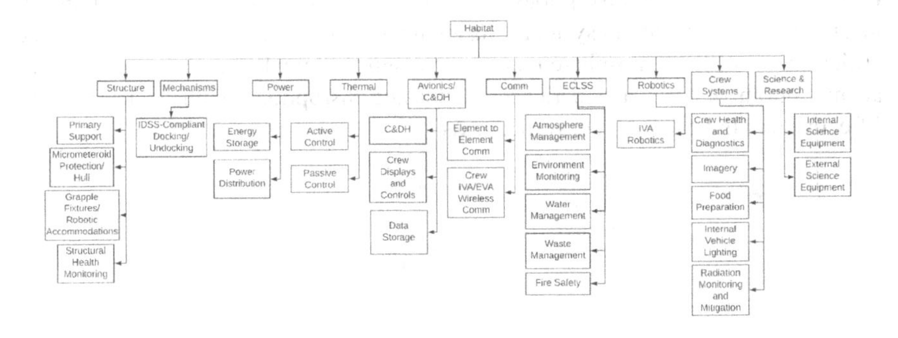

Figure 10.19: Decomposition of the 10 systems of Gateway habitation module into 27 subsystems.

atmosphere management subsystem. A few other subsystems are also highly critical.

System of Systems Modeling and Analysis often needs to process and represent a large amount of information. Figure 10.21 shows the Disruption Impact Matrix (DIM) for the Gateway habitation module. This matrix shows at a glance the impact of failures caused by subsystems on the rows to subsystems on the columns, for levels of disruptions input by the user.

Full results of this study can be found in [83].

10.5.2 INCLUDING OTHER ASPECTS OF ROPE: BUDGET AND POLICIES FOR TECHNOLOGY PRIORITIZATION

The previous examples showed analysis of space SoS mostly oriented towards characterization and modeling of the technical, physical, and operational aspects of the architectures. However, economics and policies play a very important role in SoS, and need to receive proper consideration.

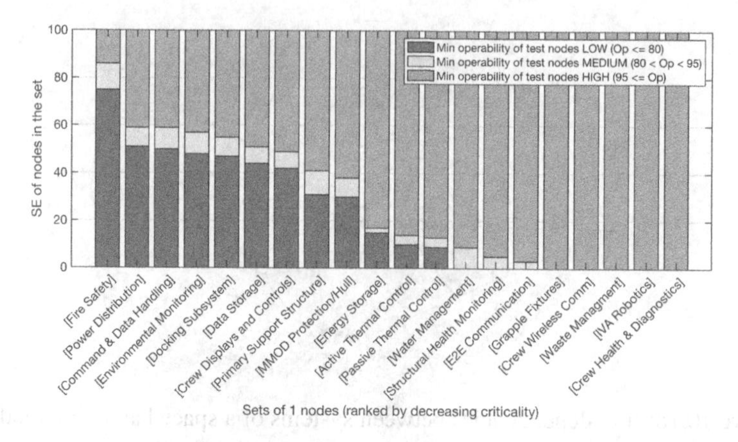

Figure 10.20: Impact of disruptions in the subsystems of Gateway habitat on the atmosphere management subsystem. Shades indicate the nominal, sub-nominal, and critical status of the atmosphere management subsystem when the system indicated at the bottom of the bar experiences increasing disruptions, with values of Self-Effectiveness indicated on the vertical axis.

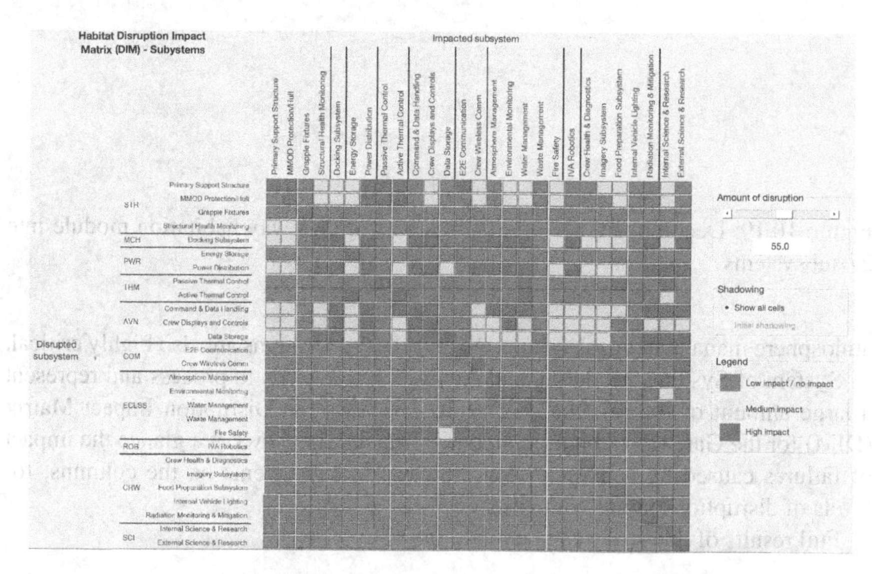

Figure 10.21: Disruption Impact Matrix for the Gateway habitation module.

Systems Developmental Dependency Analysis (Subsection 7.1.4) can be expanded to add considerations about budget limits and stakeholder preferences to the analysis of developmental dependencies. For example, we can consider technologies for Cryogenic Fluid Management (CFM). From the definition phase, Table 10.11 lists the technologies considered in this study, their association with NASA TABS categories [133], current and required Technology Readiness Level (TRL), time and cost of development, and missions enabled and supported by each technology.

Based on the developmental dependencies identified in the abstraction phase (Figure 10.22) and on the SDDA model built in the implementation phase, we can identify the expected development schedule, shown in Figure 10.23.

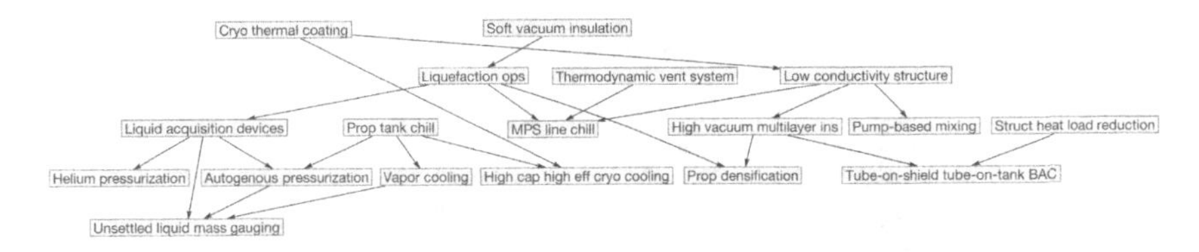

Figure 10.22: SDDA network for the CFM technologies, showing their developmental dependencies.

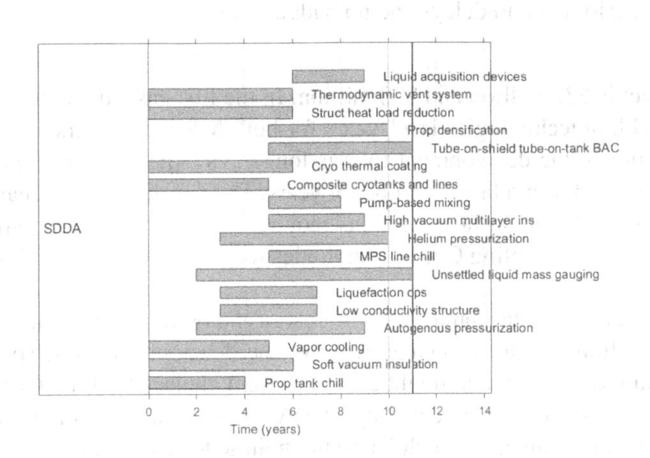

Figure 10.23: Schedule of development of CFM technologies based on SDDA model of dependencies. The light vertical line indicates when all technologies that enable lunar surface missions are available. The dark vertical line indicates when all technologies that enable Mars surface missions are available.

The model is at this point ready to be expanded to include economics and policies considerations, derived in the definition phase. For example, we can identify the yearly cost of development of CFM technologies based on the individual cost and the schedule generated by SDDA. Figure 10.24 is a sand chart showing the sum of the yearly costs of each CFM technology. The highest yearly cost is \$37.5 million.

With this extended model, the user can perform further evaluation. For example, it is interesting to assess how much longer the development will take if a lower yearly budget is available, while still preserving the developmental dependencies. Figure 10.25 shows the development of the CFM technologies when the maximum available

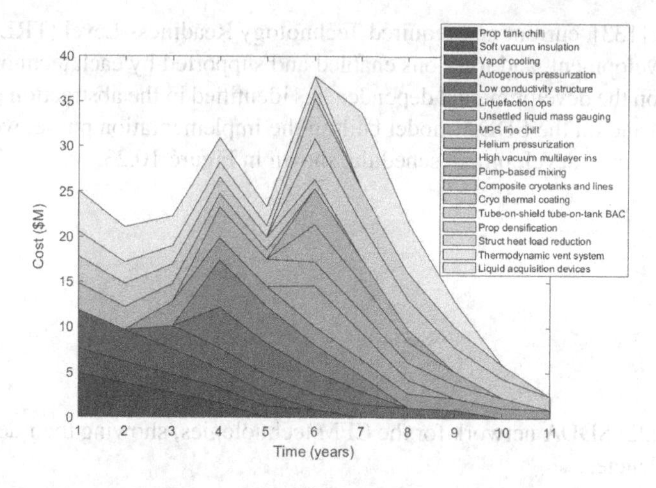

Figure 10.24: Sand chart of the cost per year to develop CFM technologies in the baseline scenario, with no delays and no budget limit.

yearly budget is $32 million and $28 million. In the first case, the time necessary to develop enabling technologies is 11 years for both Moon and Mars missions. With the lower budget, the development time is longer, as expected, since the total cost needs to be spread over a larger number of years. The development of enabling CFM technologies for Moon surface missions requires 12 years for development, while the development of enabling CFM technologies for crewed missions to Mars surface requires 13 years.

Figure 10.26 shows the sand charts of cost of the optimal development schedule under budget limits. A lower budget requires the development to be spread over a longer amount of time, which might cause different choices for initial technologies, based on the dependencies modeled by SDDA. This could in turn impact the flexibility of mission architectures built based upon these technologies.

Finally, we can use the implemented models to assess the decision-making in technology prioritization when stakeholders impose preferred technologies. Based on SDDA model, stakeholder preferences impact not only the preferred technologies, but also those necessary for the development of these preferred technologies. Compatibly with the developmental dependencies and with cost and budget limits, we want to prioritize the technologies that enable missions that are more valuable to the stakeholder. To do so, we assign a weight to express preference on technologies or missions and identify feasible development schedule (in terms of budget and developmental dependencies) that minimize the weighted completion time of the technologies. Figure 10.27 shows the different development schedules when the user prioritizes different technologies, with a yearly budget limit of $30 million and no expected delays.

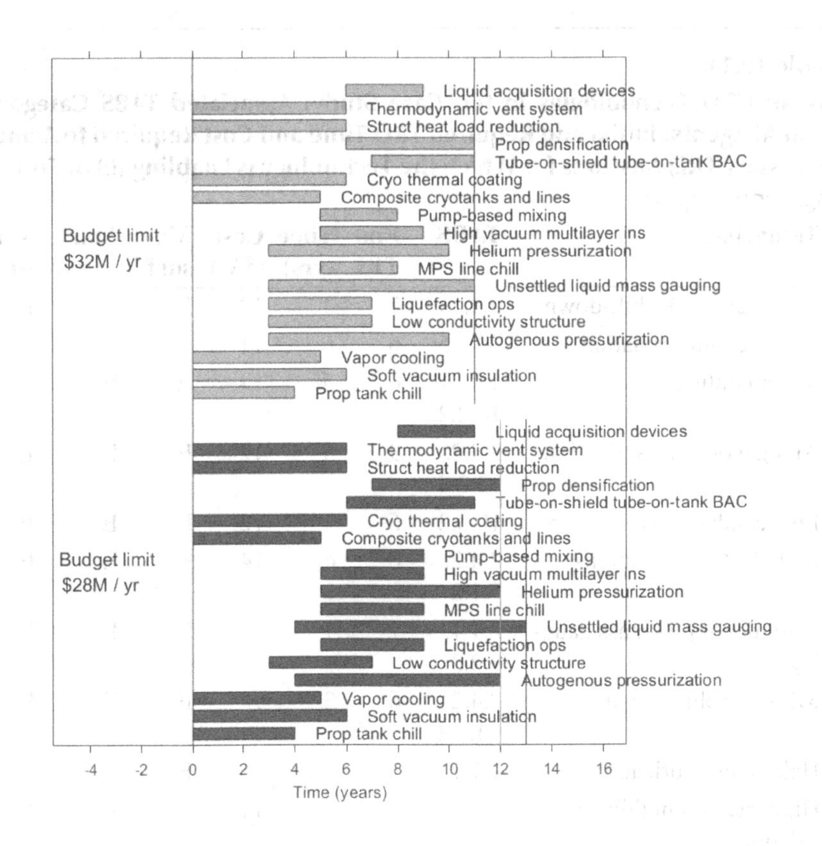

Figure 10.25: Optimal schedule of development of CFM technologies in presence of yearly budget limits. Top: $32 million limit. Bottom: $28 million limit. The dark line indicates the availability of all the enabling technologies for Mars surface. The light line indicates the availability of all the enabling technologies for lunar surface (the lines overlap in the top chart).

Table 10.11

List of CFM Technologies in the Case Study; Associated TABS Categories from AI Agents; Initial and Required TRL; Time and Cost Required to Achieve Necessary TRL; Missions for Which the Technology is Enabling (E) or Enhancing/Improving (I);

Technology	TABS	End TRL	Time (yrs)	Cost ($M)	Moon surf	Mars moons	Mars surf
Propellant tank chill-down	2.4.2	7	4	13	E	E	E
Soft vacuum insulation	10.1.5	6	6	11	-	E	E
Vapor cooling	14.1.1, 14.2.2	6	5	14	E	E	I
Autogenous pressurization	2.4.2, 12.2.3	9	5	12	E	E	E
Low conductivity structure	10.3.2	6	4	12	I	E	E
Liquefaction operations	2.4.2, 7.1.2	6	4	14	E	E	E
Unsettled liquid mass gauging	2.1.1, 9.2.3	7	7	16	I	I	E
MPS line chill-down	2.4.2, 14.1.2	6	3	9	E	E	E
Helium pressurization	2.1.1	7	7	16	E	E	E
High vacuum multilayer insulation	7.1.2	6	4	11	-	I	I
Pump-based mixing	2.1.2	6	3	8	-	E	E
Composite cryotanks and lines	12.2.5	6	5	10	I	E	E
Cryo thermal coating	14.2.2	7	6	12	E	E	I
Tube-on-shield tube-on-tank BAC	14.1.1	6	4	18	I	I	I
Prop densification	2.1.1, 2.4.2, 14.1.2	7	5	15	E	E	E
Struct heat load reduction	14.2.2	7	6	11	E	E	I
Thermodynamic vent system	2.2.3, 2.4.2	8	6	18	E	E	E
Liquid acquisition devices	2.1.2, 2.4.2, 14.1.2	7	3	17	E	E	E

MPS: Main Propulsion Systems; BAC: Broad Area Cooling

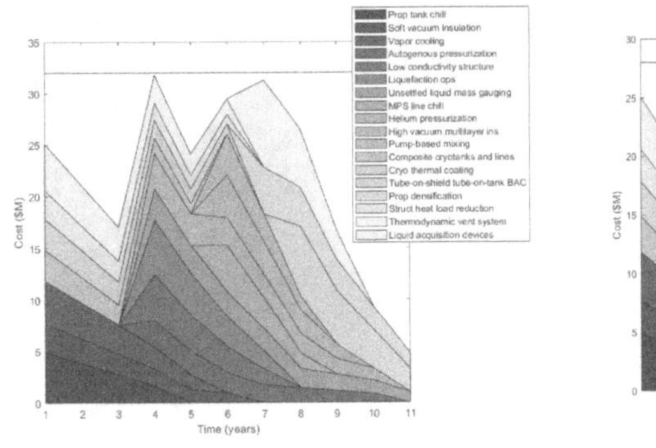

(a) $32 million yearly budget limit.

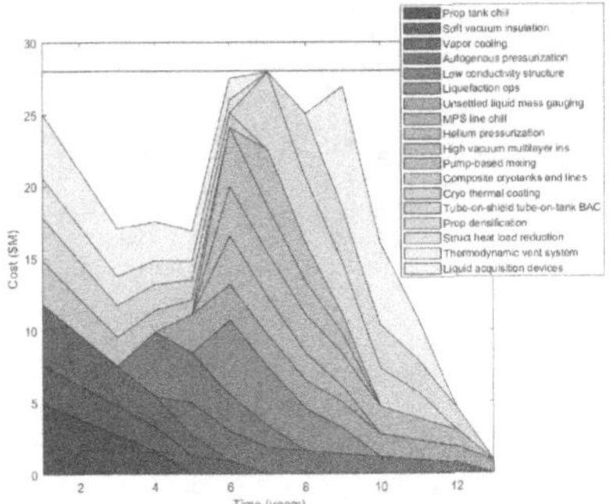

(b) $28 million yearly budget limit.

Figure 10.26: Sand chart of the yearly cost of development of CFM technologies in presence of yearly budget limits. The horizontal black line indicates the yearly budget limit. Smaller budget results in longer time to develop all the required technologies.

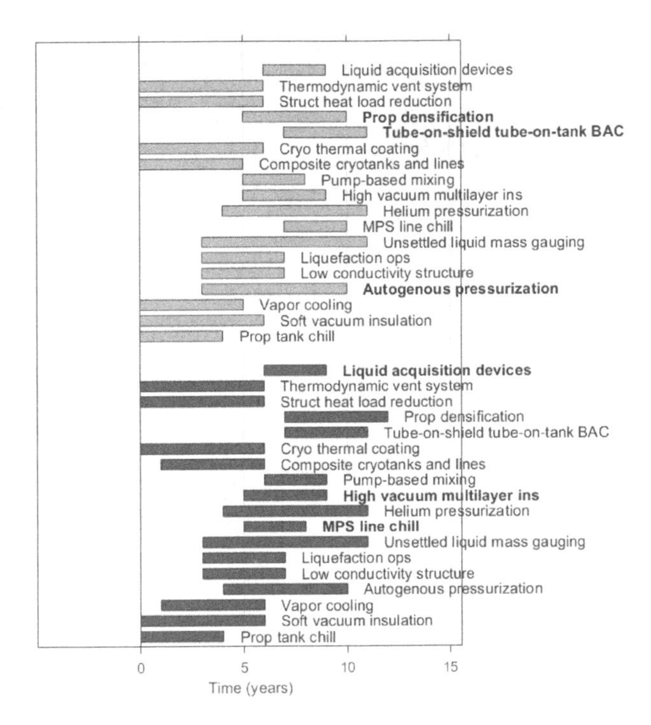

Figure 10.27: Optimal schedule of development of CFM technologies in presence of yearly budget of $30 million and no delays, with stakeholder preferences. The technologies in bold face are those with higher priority.

Glossary

AAM: Advanced Air Mobility. A range of new transformative aviation technologies.

ABM: Agent-Based Modeling. A modeling approach where each system is modeled as an independent entity.

ATS: Air Transportation System. Transportation industry comprising of airlines and their supporting industries and infrastructure.

AWB: Analytic Workbench. Suite of tools for SoS analysis

- **MUSTDO:** Multi-Stakeholder Dynamic Optimization.
- **RPO:** Robust Portfolio Optimization.
- **SDDA:** Systems Developmental Dependency Analysis.
- **SODA:** Systems Operational Dependency Analysis.

BKI: Behavior-Knowledge-Intent. A model of agent used in agent-based modeling.

DAI: Definition-Abstraction-Implementation. A three-phase approach to modeling and analysis of systems-of-systems.

PSU: Provider of Services for UAM. Airspace service provider responsible for providing air traffic management services for the urban air mobility market.

RAM: Regional Air Mobility.

SE: Systems Engineering.

SoS: System-of-Systems. A type of complex system whose constituents are themselves independent systems.

UAM: Urban Air Mobility. A subset of advanced air mobility market comprising air transportation services in urban areas.

UAS: Unmanned Aerial System.

UTM: UAS Traffic Management.

References

1. System modeling language (sysml) project. Available at http://sysml.org (2022/02/03).
2. ISO/IEC/IEEE 15288:2015. Systems and software engineering — system life cycle processes. Standard, International Organization for Standardization, Geneva, CH, May 2015.
3. ISO/IEC/IEEE 21839:2019. Systems and software engineering – System of systems (SoS) considerations in life cycle stages of a system. Standard, International Organization for Standardization, Geneva, CH, July 2019.
4. ISO/IEC/IEEE 21840:2019. Systems and software engineering – guidelines for the utilization of iso/iec/ieee 15288 in the context of system of systems (sos). Standard, International Organization for Standardization, Geneva, CH, December 2019.
5. ISO/IEC/IEEE 21841:2019. Systems and software engineering — taxonomy of systems of systems. Standard, International Organization for Standardization, Geneva, CH, July 2019.
6. Russ Abbott. Open at the top; open at the bottom; and continually (but slowly) evolving. In *2006 IEEE/SMC International Conference on System of Systems Engineering*, pages 6–pp. IEEE, 2006.
7. Russell L. Ackoff. Towards a System of Systems Concepts. *Management Science*, 17(11):661–671, 1971. Publisher: INFORMS.
8. Kevin Macgregor Adams, Patrick T. Hester, and Joseph M. Bradley. A Historical Perspective of Systems Theory. In *Proceedings of the 2013 Industrial and Systems Engineering Research Conference*, San Juan, Puerto Rico, May 2013.
9. Shafeeq Ahmad and Vipin Saxena. Design of formal air traffic control system through uml. *Ubiquitous computing and communication journal*, 3(6):11–20, 2008.
10. Réka Albert and Albert-László Barabási. Statistical mechanics of complex networks. *Reviews of modern physics*, 74(1):47, 2002.
11. James T. Allison. Complex System Optimization: A Review of Analytical Target Cascading, Collaborative Optimization, and Other Formulations. Master's thesis, Department of Mechanical Engineering, University of Michigan, Ann Arbor, 2004.
12. Robert Axtell. Why agents?: on the varied motivations for agent computing in the social sciences. In *Proceedings of Agent Simulation: Applications, Models, and Tools*. Center on Social and Economic Dynamics Brookings Institution, 2000.
13. Carliss Baldwin, Alan MacCormack, and John Rusnak. Hidden structure: Using network methods to map system architecture. *Research Policy*, 43(8):1381–1397, October 2014.
14. W. Clifton Baldwin and Wilson N. Felder. Mathematical Characterization of System-of-Systems Attributes. In Franz-Josef Kahlen, Shannon Flumerfelt, and Anabela Alves, editors, *Transdisciplinary Perspectives on Complex Systems*, pages 1–24. Springer International Publishing, Cham, 2017.
15. Albert-László Barabási and Márton Pósfai. *Network science*. Cambridge University Press, Cambridge, United Kingdom, 2016. OCLC: ocn910772793.
16. J. Clark Beesemyer, Adam M. Ross, and Donna H. Rhodes. An empirical investigation of system changes to frame links between design decisions and ilities. *Procedia Computer Science*, 8:31–38, 2012.

17. Peter Belobaba, Amedeo Odoni, and Cynthia Barnhart, editors. *The Global Airline Industry*. Wiley, 1 edition, April 2009.
18. Aharon Ben-Tal and Arkadi Nemirovski. Robust solutions of linear programming problems contaminated with uncertain data. *Mathematical programming*, 88(3):411–424, 2000.
19. Brian J. L. Berry. Cities as systems within systems of cities. *Papers of the Regional Science Association*, 13(1):146–163, December 1964.
20. Dimitri P Bertsekas and John N Tsitsiklis. Neuro-dynamic programming: an overview. In *Proceedings of 1995 34th IEEE conference on decision and control*, volume 1, pages 560–564. IEEE, 1995.
21. Benjamin S. Blanchard and W. J. Fabrycky. *Systems engineering and analysis*. Prentice Hall international series in industrial and systems engineering. Prentice Hall, Boston, 5th ed edition, 2011. OCLC: ocn457159070.
22. Benjamin S Blanchard and Wolter J Fabrycky. Systems engineering and analysis. 2006.
23. Lawrence Blume. Agent-based models for policy analysis. In *Assessing the Use of Agent-Based Models for Tobacco Regulation*. National Academies Press (US), 2015.
24. SEBoK Editorial Board. *The Guide to the Systems Engineering Body of Knowledge (SEBoK)*. Stevens Institute of Technology Systems Engineering Research Center, Hoboken, NJ, 2.3 edition, 2020.
25. J. Boardman and B. Sauser. System of Systems - the meaning of of. In *2006 IEEE/SMC International Conference on System of Systems Engineering*, pages 118–123, Los Angeles, California, USA, 2006. IEEE.
26. Barry Boehm, Jo Ann Lane, Supannika Koolmanojwong, and Richard Turner. *The incremental commitment spiral model: Principles and practices for successful systems and software*. Addison-Wesley Professional, 2014.
27. E. Bonabeau. Agent-based modeling: Methods and techniques for simulating human systems. *Proceedings of the National Academy of Sciences*, 99(Supplement 3):7280–7287, May 2002.
28. Kevin H. Bonanne. A model-based approach to System-of-Systems engineering via the systems modeling language. Master's thesis, Purdue University, 2014.
29. Kenneth E Boulding. General Systems Theory - The Skeleton Of Science. *Management science*, 2(3):197–208, April 1956. INFORMS.
30. George E. P. Box, J. Stuart Hunter, and William Gordon Hunter. *Statistics for experimenters: design, innovation, and discovery*. Wiley-Interscience, 2 edition, 2005.
31. Dan Braha. The Complexity of Design Networks: Structure and Dynamics. In Philip Cash, Tino Stanković, and Mario Štorga, editors, *Experimental Design Research*, pages 129–151. Springer International Publishing, Cham, 2016.
32. T. R. Browning. Applying the Design Structure Matrix to System Decomposition and Integration Problems: a Review and New Directions. *IEEE Transactions on Engineering Management*, 48(3):292–306, 2001.
33. Stuart E. Burge. Systems Engineering: Using Systems Thinking to Design Better Aerospace Systems. In Richard Blockley and Wei Shyy, editors, *Encyclopedia of Aerospace Engineering*. John Wiley & Sons, Ltd, Chichester, UK, December 2010.
34. Arianne X. Collopy, Eytan Adar, and Panos Y. Papalambros. On the use of coordination strategies in complex engineered system design projects. *Design Science*, 6:e32, 2020.
35. Committee on Enhancing Air Mobility—A National Blueprint, Aeronautics and Space Engineering Board, Division on Engineering and Physical Sciences, and National Academies of Sciences, Engineering, and Medicine. *Advanced Aerial Mobility: A National Blueprint*. National Academies Press, Washington, D.C., 2020.

36. Gerard Cornuejols and Reha Tütüncü. *Optimization methods in finance*, volume 5. Cambridge University Press, 2006.
37. John W Creswell and J David Creswell. *Research design: Qualitative, quantitative, and mixed methods approaches*. Sage publications, 2017.
38. Kenneth L Cureton and F Stan Settles. Systems-of-systems architecting: educational findings and implications. In *2005 IEEE International Conference on Systems, Man and Cybernetics*, volume 3, pages 2726–2731. IEEE, 2005.
39. Adam Dachowicz, Kshitij Mall, Prajwal Balasubramani, Apoorv Maheshwari, Ali K Raz, Jitesh H Panchal, and Daniel DeLaurentis. Mission engineering and design using real-time strategy games: An explainable AI approach. *Journal of Mechanical Design*, 144(2), 2022.
40. J Dahmann, G Rebovich, J A Lane, and R Lowry. System engineering artifacts for SoS. *IEEE Aerospace and Electronic Systems Magazine*, 26(1):22–28, January 2011.
41. Judith Dahmann, George Rebovich, JoAnn Lane, Ralph Lowry, and Kristen Baldwin. An implementers' view of systems engineering for systems of systems. In *Systems Conference (SysCon), 2011 IEEE International*, pages 212–217. IEEE, 2011.
42. Navindran Davendralingam and Daniel DeLaurentis. A robust optimization framework to architecting system of systems. *Procedia Computer Science*, 16:255–264, 2013.
43. Navindran Davendralingam and Daniel DeLaurentis. An analytic portfolio approach to system of systems evolutions. *Procedia Computer Science*, 28:711–719, 2014.
44. Navindran Davendralingam and Daniel DeLaurentis. A robust portfolio optimization approach to system of system architectures. *Systems Engineering*, 18(3):269–283, 2015.
45. Daniel DeLaurentis. Understanding transportation as a system-of-systems design problem. In *43rd AIAA Aerospace Sciences Meeting and Exhibit*, volume 1, pages 10–13. Reno, NV New York, NY, 2005.
46. Daniel DeLaurentis, Robert K Callaway, et al. A system-of-systems perspective for public policy decisions. *Review of Policy Research*, 21(6):829–837, 2004.
47. Daniel DeLaurentis, William A. Crossley, and Muharrem Mane. Taxonomy to Guide Systems-of-Systems Decision-Making in Air Transportation Problems. *Journal of Aircraft*, 48(3):760–770, May 2011.
48. Daniel DeLaurentis, E.-P. Han, and Tatsuya Kotegawa. Network-theoretic approach for analyzing connectivity in air transportation networks. *Journal of Aircraft*, 45(5):1669–1679, 2008.
49. Daniel DeLaurentis, Jitesh H Panchal, Ali K Raz, Prajwal Balasubramani, Apoorv Maheshwari, Adam Dachowicz, and Kshitij Mall. Toward automated game balance: A systematic engineering design approach. In *2021 IEEE Conference on Games (CoG)*, pages 1–8. IEEE, 2021.
50. JE Dennis, Sharon F Arroyo, Evin J Cramer, and Paul D Frank. Problem formulations for systems of systems. In *2005 IEEE International Conference on Systems, Man and Cybernetics*, volume 1, pages 64–71. IEEE, 2005.
51. Homayoon Dezfuli. Achieving a holistic and risk-informed decision-making process at nasa. In *Workshop on Tolerable Risk Evaluation*, Arlington, VA, 2008.
52. DoD. Department of defense architecture framework (DODAF).
53. Dov Dori, Hillary Sillitto, Regina M. Griego, Dorothy McKinney, Eileen P. Arnold, Patrick Godfrey, James Martin, Scott Jackson, and Daniel Krob. System Definition, System Worldviews, and Systemness Characteristics. *IEEE Systems Journal*, 14(2):1538–1548, June 2020.

54. Allen Downey. *Think complexity: complexity science and computational modeling*. O'Reilly, Beijing ; Boston, second edition, 2018. OCLC: on1049571662.
55. H. Eisner, J. Marciniak, and R. McMillan. Computer-aided system of systems (S2) engineering. In *Conference Proceedings 1991 IEEE International Conference on Systems, Man, and Cybernetics*, pages 531–537, Charlottesville, VA, USA, 1991. IEEE.
56. Steven D Eppinger and Tyson R Browning. *Design structure matrix methods and applications*. MIT press, 2012.
57. Jörn Ewaldt. A System Dynamics Analysis of the Effects of Capacity Limitations in a Multi-Level Production Chain. Bergen, Norway, August 2000.
58. Kristopher L Ezra, Daniel DeLaurentis, Linas Mockus, and Joseph F Pekny. Developing mathematical formulations for the integrated problem of sensors, weapons, and targets. *Journal of Aerospace Information Systems*, 13(5):175–190, 2016.
59. Samuel F. Kovacic, Andres Sousa-Poza, and Charles Keating. The National Centers for System of Systems Engineering: A Case Study on Shifting the Paradigm for System of Systems. *Systems Research Forum*, 02(01):52–58, January 2007.
60. Frank J Fabozzi, Sergio M Focardi, Petter N Kolm, and Dessislava A Pachamanova. *Robust portfolio optimization and management*. John Wiley & Sons, 2007.
61. Zhemei Fang, Navindran Davendralingam, and Daniel DeLaurentis. Multistakeholder dynamic optimization for acknowledged system-of-systems architecture selection. *IEEE Systems Journal*, 12(4):3565–3576, 2018.
62. Federal Aviation Administration (FAA). Urban air mobility (uam) concept of operations (conops) v1.0. 2020.
63. Ronald Fisher. The arrangement of field experiments. *Journal of the Ministry of Agriculture of Great Britain*, 33:503–513, 1926.
64. Jay W. Forrester. Lessons from system dynamics modeling. *System Dynamics Review*, 3(2):136–149, 1987.
65. Jay W. Forrester. System dynamics, systems thinking, and soft OR. *System Dynamics Review*, 10(2-3):245–256, 1994.
66. Sanford Friedenthal, Regina Griego, and Mark Sampson. Incose model based systems engineering (mbse) initiative. In *INCOSE 2007 symposium*, volume 11. sn, 2007.
67. Henry Laurence Gantt. *Work, wages, and profits: their influence on the cost of living*. Engineering magazine, 1910.
68. Martin Gardner. Mathematical Games: The fantastic combinations of john conway's new solitaire game "life". *Scientific American*, 223(4):120–123, October 1970.
69. Paul Garvey and Ariel Pinto. Introduction to functional dependency network analysis. In *Second International Engineering Systems Symposium*, Cambridge, Massachusetts, 15-17 June 2009.
70. Paul Garvey and Ariel Pinto. *Advanced Risk Analysis in Engineering Enterprise Systems*. CRC Press, 2012.
71. Paul R Garvey, C Ariel Pinto, and Joost Reyes Santos. Modelling and measuring the operability of interdependent systems and systems of systems: advances in methods and applications. *International Journal of System of Systems Engineering*, 5(1):1–24, 2014.
72. Murray Gell-Mann. *The Quark and the Jaguar: Adventures in the Simple and the Complex*. Macmillan, 1995.
73. Ronald E Giachetti. Evaluation of the dodaf meta-model's support of systems engineering. *Procedia Computer Science*, 61:254–260, 2015.
74. John E Gibson, William T Scherer, William F Gibson, and Wiley InterScience (Online service). *How to Do systems analysis*. Wiley-Interscience, Hoboken, N.J., 2007. OCLC: 181345418.

75. G. Nigel Gilbert. *Agent-based models*. SAGE Publications, Thousand Oaks, California, 2nd edition edition, 2020.
76. Alex Gorod, Brian Sauser, and John Boardman. System-of-systems engineering management: A review of modern history and a path forward. *IEEE Systems Journal*, 2(4):484–499, 2008.
77. Rohit Goyal, Colleen Reiche, Chris Fernando, Jacquie Serrao, Shawn Kimmel, Adam Cohen, and Susan Shaheen. Urban air mobility (uam) market study. Technical report, 2018.
78. US GPO. Restructuring of the strategic defense initiative (sdi) program. In *Services, USCS C. o. A.(Ed.), United States Congress*, page 16, 1989.
79. Melissa T Greene and Panos Y Papalambros. A cognitive framework for engineering systems thinking. In *2016 Conference on Systems Engineering Research*, 2016.
80. Cesare Guariniello. *Supporting space systems design via systems dependency analysis methodology*. PhD thesis, Purdue University, 2016.
81. Cesare Guariniello and Daniel DeLaurentis. Dependency analysis of system-of-systems operational and development networks. *Procedia Computer Science*, 16:265–274, 2013.
82. Cesare Guariniello, Zhemei Fang, Navindran Davendralingam, Karen Marais, and Daniel DeLaurentis. Tool suite to support model based systems engineering-enabled system-of-systems analysis. In *2018 IEEE Aerospace Conference*, pages 1–16. IEEE, 2018.
83. Cesare Guariniello, Melanie Grande, Christopher Brand, Liam Durbin, Michael Dai, Ashwati Das-Stuart, Reginald A Alexander, Kathleen Howell, and Daniel DeLaurentis. Quantifying the impact of systems interdependencies in space systems architectures. In *International Astronautical Congress (IAC) 2019*, number MSFC-E-DAA-TN74200, 2019.
84. Cesare Guariniello, Thomas B Marsh, Thomas Diggelmann, and Daniel DeLaurentis. System-of-systems methods for technology assessment and prioritization for space architectures. In *2021 IEEE Aerospace Conference (50100)*, pages 1–13. IEEE, 2021.
85. Cesare Guariniello, Linas Mockus, Ali K Raz, and Daniel DeLaurentis. Towards intelligent architecting of aerospace system-of-systems. In *2019 IEEE Aerospace Conference*, pages 1–11. IEEE, 2019.
86. Cesare Guariniello, Linas Mockus, Ali K Raz, and Daniel DeLaurentis. Towards intelligent architecting of aerospace system-of-systems: Part ii. In *2020 IEEE Aerospace Conference*, pages 1–9. IEEE, 2020.
87. Cesare Guariniello, William O'Neill, Ashwati Das-Stuart, Liam Durbin, Kathleen Howell, Reginald A Alexander, and Daniel DeLaurentis. System-of-systems tools for the analysis of technological choices in space propulsion. In *International Astronautical Congress*, number IAC-18, 2018.
88. John Guckenheimer and Julio M Ottino. Foundations for Complex Systems Research in the Physical Sciences and Engineering. Technical report, National Science Foundation, 2008.
89. David Gunning and David Aha. Darpa's explainable artificial intelligence (xai) program. *AI Magazine*, 40(2):44–58, 2019.
90. Davis Hackenberg. Nasa grand challenge update for vfs. In *Annual Vertical Flight Society (VFS) Forum and Technology Display (Forum 75)*, number AFRC-E-DAA-TN68883, 2019.

91. Yacov Y Haimes, Barry M Horowitz, James H Lambert, Joost R Santos, Chenyang Lian, and Kenneth G Crowther. Inoperability input-output model for interdependent infrastructure sectors. i: Theory and methodology. *Journal of Infrastructure Systems*, 11(2):67–79, 2005.
92. Yacov Y Haimes and Pu Jiang. Leontief-based model of risk in complex interconnected infrastructures. *Journal of Infrastructure systems*, 7(1):1–12, 2001.
93. Reiko Heckel, Alexander Kurz, and Edmund Chattoe-Brown. Features of Agent-based Models. *Electronic Proceedings in Theoretical Computer Science*, 263:31–37, December 2017.
94. George H. Heilmeier, "The Heilmeier Catechism," Defense Advanced Research Projects Agency (DARPA), https://www.darpa.mil/work-with-us/heilmeier-catechism.
95. Brian P Hill, Dwight DeCarme, Matt Metcalfe, Christine Griffin, Sterling Wiggins, Chris Metts, Bill Bastedo, Michael D Patterson, and Nancy L Mendonca. Uam vision concept of operations (conops) uam maturity level (uml) 4. 2020.
96. Jack Hirshleifer. On the economics of transfer pricing. *The Journal of Business*, 29(3):172–184, 1956.
97. Steven R Hirshorn, Linda D Voss, and Linda K Bromley. *NASA Systems Engineering Handbook Revision 2*. Revision 2 edition, 2017.
98. Oliver Hoehne. The sos-vee model: Mastering the socio-technical aspects and complexity of systems of systems engineering (sose). In *INCOSE International Symposium*, volume 26, pages 1494–1508. Wiley Online Library, 2016.
99. John H Holland. *Hidden order: How adaptation builds complexity*. Addison Wesley Longman Publishing Co., Inc., 1996.
100. John H. Holland. Studying Complex Adaptive Systems. *Journal of Systems Science and Complexity*, 19(1):1–8, March 2006.
101. Steven Holt, Paul Collopy, and Dianne DeTurris. So It's Complex, Why Do I Care? In Franz-Josef Kahlen, Shannon Flumerfelt, and Anabela Alves, editors, *Transdisciplinary Perspectives on Complex Systems*, pages 25–48. Springer International Publishing, Cham, 2017.
102. Ryan S. Hutcheson, Daniel McAdams, Robert B. Stone, and Irem Y. Tumer. Function-based behavioral modeling. In *ASME International Design Engineering Technical Conferences and Computers and Information in Engineering Conference*, pages 547–558. American Society of Mechanical Engineers, 2007.
103. Thomas V Huynh and John S Osmundson. An integrated systems engineering methodology for analyzing systems of systems architectures. In *Asia-Pacific Systems Engineering Conference, Singapore*, page 77, 2007.
104. ICAO. Global tbo concept - draft material in development by an icao expert group. Technical report, International Civil Aviation Organization.
105. INCOSE. *Systems Engineering Handbook: A Guide for System Life Cycle Processes and Activities*. John Wiley & Sons, 2015.
106. INCOSE-TP-2018-003-01.0. Systems of systems primer. Technical report, International Council on Systems Engineering, San Diego, CA, 2018.
107. Mo Jamshidi. *Systems of systems engineering: principles and applications*. CRC press, 2017.
108. Mohammad Jamshidi and Andrew P Sage. *System of systems engineering: innovations for the 21st century*, volume 58. John Wiley & Sons Incorporated, 2009.
109. Jerzy Kamburowski. Normally distributed activity durations in PERT networks. *Journal of the Operational Research Society*, 36(11):1051–1057, 1985.

110. Michael Kokkolaras, Zissimos P Mourelatos, and Panos Y Papalambros. Design optimization of hierarchically decomposed multilevel systems under uncertainty. In *International Design Engineering Technical Conferences and Computers and Information in Engineering Conference*, volume 46946, pages 613–624, 2004.
111. Tatsuya Kotegawa, Daniel A DeLaurentis, and Aaron Sengstacken. Development of network restructuring models for improved air traffic forecasts. *Transportation Research Part C: Emerging Technologies*, 18(6):937–949, 2010.
112. Vadim Kotov. Systems of systems as communicating structures, 1997.
113. Erwin Kreyszig. *Advanced engineering mathematics*. John Wiley & Sons, 2010.
114. Viswanathan Krishnan, Steven D. Eppinger, and Daniel E. Whitney. A Model-Based Framework to Overlap Product Development Activities. *Management Science*, 43(4):437–451, 1997.
115. Jarret M Lafleur. *A markovian state-space framework for integrating flexibility into space system design decisions*. PhD thesis, Georgia Institute of Technology, 2011.
116. Robert J. Lempert, Steven W. Popper, and Steven C. Bankes. *Shaping the Next One Hundred Years: New Methods for Quantitative, Long-Term Policy Analysis*. RAND Corporation, 2003.
117. Wassily W Leontief. Quantitative input and output relations in the economic systems of the united states. *The review of economic statistics*, 18(3):105–125, 1936.
118. JH Lewe, Daniel DeLaurentis, and Dimitri N Mavris. Foundation for study of future transportation systems through agent-based simulation. In *24th International Congress of the Aeronautical Sciences*, volume 29, 2004.
119. C.M. Macal and M.J. North. Tutorial on agent-based modeling and simulation. In *Proceedings of the Winter Simulation Conference, 2005.*, pages 2–15, Orlando, FL, USA, 2005. IEEE.
120. Azad M Madni and Michael Sievers. System of systems integration: Fundamental concepts, challenges and opportunities. *Advances in Systems Engineering, American Institute of Aeronautics and Astronautics, Reston, VA*, pages 1–34, 2016.
121. Mark W Maier. Architecting principles for systems-of-systems. *Systems Engineering*, 1(4):267–274, 1998.
122. Mark W. Maier and Eberhardt Rechtin. *The art of systems architecting*. CRC Press, Boca Raton, 3rd ed edition, 2009. OCLC: 226357386.
123. Muharrem Mane and Daniel DeLaurentis. Sensor platform management strategies in a multi-threat environment. In *Infotech@ Aerospace 2012*, page 2546. 2012.
124. Harry Markowitz. Portfolio selection. *The Journal of Finance*, 7(1):77–91, 1952.
125. Gautam Marwaha and Michael Kokkolaras. System-of-systems approach to air transportation design using nested optimization and direct search. *Structural and Multidisciplinary Optimization*, 51(4):885–901, 2015.
126. Peter McBurney. What Are Models for? In David Hutchison, Takeo Kanade, Josef Kittler, Jon M. Kleinberg, Friedemann Mattern, John C. Mitchell, Moni Naor, Oscar Nierstrasz, C. Pandu Rangan, Bernhard Steffen, Madhu Sudan, Demetri Terzopoulos, Doug Tygar, Moshe Y. Vardi, Gerhard Weikum, Massimo Cossentino, Michael Kaisers, Karl Tuyls, and Gerhard Weiss, editors, *Multi-Agent Systems*, volume 7541, pages 175–188. Springer Berlin Heidelberg, Berlin, Heidelberg, 2012. Series Title: Lecture Notes in Computer Science.
127. Richard Meltzer. Basics of Modeling: What, Why, and How. In *NECSI Summer School*, pages 6–pp. New England Complexity Sciences Institute, 2005.
128. MITRE. *Systems Engineering Guide*. The MITRE Corporation, 2014.

129. Douglas C Montgomery, George C Runger, and Norma F Hubele. *Engineering statistics*. John Wiley & Sons, 2009.
130. Kushal Moolchandani, Datu Buyung Agusdinata, Muharrem Mane, Daniel DeLaurentis, and William Crossley. Assessment of the effect of aircraft technological advancement on aviation environmental impacts. In *51st AIAA Aerospace Sciences Meeting including the New Horizons Forum and Aerospace Exposition*, page 652, 2013.
131. Kushal Moolchandani, Parithi Govindaraju, Satadru Roy, William A. Crossley, and Daniel DeLaurentis. Assessing Effects of Aircraft and Fuel Technology Advancement on Select Aviation Environmental Impacts. *Journal of Aircraft*, 54(3):857–869, May 2017.
132. Kushal Moolchandani, Zhenghui Sha, Apoorv Maheshwari, Joseph Thekinen, Navindran Davendralingam, Jitesh Panchal, and Daniel A. DeLaurentis. Hierarchical Decision-Modeling Framework for Air Transportation System. In *16th AIAA Aviation Technology, Integration, and Operations Conference*, Washington, D.C., June 2016. American Institute of Aeronautics and Astronautics.
133. NASA. Nasa technology roadmaps archive, 2015. www.nasa.gov/offices/oct/home/roadmaps/index.html, accessed 2022-03-07.
134. Kartavya Neema, Shreyas Vathul Subramanian, and Daniel Delaurentis. Dual phase consensus algorithm for distributed sensor management. *IEEE Transactions on Aerospace and Electronic Systems*, 52(4):1893–1907, 2016.
135. M. E. J. Newman. Complex Systems: A Survey. *American Journal of Physics*, 79(8):800–810, August 2011. arXiv: 1112.1440.
136. Mark EJ Newman. The structure and function of complex networks. *SIAM review*, 45(2):167–256, 2003.
137. S. K. Numrich and Andreas Tolk. Challenges for Human, Social, Cultural, and Behavioral Modeling. *SCS M&S Magazine*, 4(1):1–9, 2010.
138. James Ostrowski, Miguel F Anjos, and Anthony Vannelli. Tight mixed integer linear programming formulations for the unit commitment problem. *IEEE Transactions on Power Systems*, 27(1):39–46, 2012.
139. Gregory S Parnell and Timothy E Trainor. Using the swing weight matrix to weight multiple objectives. In *INCOSE International Symposium*, volume 19, pages 283–298. Wiley Online Library, 2009.
140. Hans Polzer, Daniel DeLaurentis, and Donald N. Fry. Multiplicity of Perspectives, Context Scope, and Context Shifting Events. In *2007 IEEE International Conference on System of Systems Engineering*, pages 1–6, San Antonio, TX, USA, April 2007. IEEE.
141. Warren B Powell. *Approximate Dynamic Programming: Solving the curses of dimensionality*, volume 703. John Wiley & Sons, 2007.
142. Ali K Raz and Daniel DeLaurentis. Information fusion system design space characterization by design of experiments. In *2016 19th International Conference on Information Fusion (FUSION)*, pages 2147–2154. IEEE, 2016.
143. Ali K Raz, Paul C Wood, Linas Mockus, and Daniel DeLaurentis. System of systems uncertainty quantification using machine learning techniques with smart grid application. *Systems Engineering*, 23(6):770–782, 2020.
144. Garry Robins, Pip Pattison, Yuval Kalish, and Dean Lusher. An introduction to exponential random graph (p*) models for social networks. *Social Networks*, 29(2):173–191, May 2007.

145. Andrew P. Sage and Christopher D. Cuppan. On the Systems Engineering and Management of Systems of Systems and Federations of Systems. *Information Knowledge Systems Management*, 2(4):325–345, 2001. IOS Press.
146. Robert G Sargent. Some approaches and paradigms for verifying and validating simulation models. In *Proceeding of the 2001 Winter Simulation Conference (Cat. No. 01CH37304)*, volume 1, pages 106–114. IEEE, 2001.
147. Sarah A. Sheard and Ali Mostashari. 7.3.1 A Complexity Typology for Systems Engineering. *INCOSE International Symposium*, 20(1):933–945, July 2010.
148. Aaron Shenhar. A new systems engineering taxonomy. In *Proceedings of the 4th International Symposium of the National Council on System Engineering, National Council on System Engineering*, volume 2, pages 261–276, 1994.
149. Herbert A Simon. *The sciences of the artificial.* MIT press, 1996.
150. Joseph J Simpson and Mary J Simpson. System of systems complexity identification and control. In *2009 IEEE International Conference on System of Systems Engineering (SoSE)*, pages 1–6. IEEE, 2009.
151. Oleg V Sindiy, Daniel DeLaurentis, and William B Stein. An agent-based dynamic model for analysis of distributed space exploration architectures. *The Journal of the Astronautical Sciences*, 57(3):579–606, 2009.
152. Jaroslaw Sobieszczanski-Sobieski. Integrated system-of-systems synthesis. *AIAA journal*, 46(5):1072–1080, 2008.
153. Robert B. Stone, Irem Y. Tumer, and Michael Van Wie. The function-failure design method. *Journal of Mechanical Design*, 127(3):397–407, 2005.
154. Robert B. Stone and Kristin L. Wood. Development of a functional basis for design. *Journal of Mechanical design*, 122(4):359–370, 2000.
155. Genichi Taguchi. *Introduction to quality engineering: designing quality into products and processes*. 1986.
156. S. Tamaskar, K. Neema, T. Kotegawa, and D. DeLaurentis. Complexity enabled design space exploration. In *IEEE International Conference on Systems, Man, and Cybernetics*, pages 1250–1255, 2011.
157. Shashank Tamaskar, Kartavya Neema, and Daniel DeLaurentis. Framework for measuring complexity of aerospace systems. *Research in Engineering Design*, 25(2):125–137, 2014.
158. Zhong W Thai, Prajwal Balasubramani, Chris Brand, Andrew Haines, and Daniel DeLaurentis. Study of swarm-based planetary exploration architectures using agent-based modeling. In *AIAA Scitech 2020 Forum*, page 0075, 2020.
159. David P. Thipphavong, Rafael Apaza, Bryan Barmore, Vernol Battiste, Barbara Burian, Quang Dao, Michael Feary, Susie Go, Kenneth H. Goodrich, Jeffrey Homola, Husni R. Idris, Parimal H. Kopardekar, Joel B. Lachter, Natasha A. Neogi, Hok Kwan Ng, Rosa M. Oseguera-Lohr, Michael D. Patterson, and Savita A. Verma. Urban Air Mobility Airspace Integration Concepts and Considerations. In *2018 Aviation Technology, Integration, and Operations Conference*, Atlanta, Georgia, June 2018. American Institute of Aeronautics and Astronautics.
160. Irem Y Tumer and Robert B Stone. Mapping function to failure mode during component development. *Research in Engineering Design*, 14(1):25–33, 2003.
161. Reha H Tütüncü and M Koenig. Robust asset allocation. *Annals of Operations Research*, 132(1-4):157–187, 2004.
162. Payuna Uday and Karen Marais. Designing Resilient Systems-of-Systems: A Survey of Metrics, Methods, and Challenges. *Systems Engineering*, 18(5):491–510, 2015.

163. Stanislav Uryasev. Conditional value-at-risk: Optimization algorithms and applications. In *proceedings of the IEEE/IAFE/INFORMS 2000 conference on computational intelligence for financial engineering (CIFEr)(Cat. No. 00TH8520)*, pages 49–57. IEEE, 2000.
164. Savita A. Verma, Spencer C. Monheim, Kushal A. Moolchandani, Priyank Pradeep, Annie W. Cheng, David P. Thipphavong, Victoria L. Dulchinos, Heather Arneson, Todd A. Lauderdale, Christabelle S. Bosson, Eric R. Mueller, and Bogu Wei. Lessons Learned: Using UTM Paradigm for Urban Air Mobility Operations. In *2020 AIAA/IEEE 39th Digital Avionics Systems Conference (DASC)*, pages 1–10, San Antonio, TX, USA, October 2020. IEEE.
165. Ludwig Von Bertalanffy. *General system theory: foundations, development, applications*. George Braziller, New York, NY, USA, 1968.
166. Wernher Von Braun and Henry J White. *The Mars Project*. University of Illinois Press, 1953.
167. David D. Walden, Garry J. Roedler, Kevin Forsberg, R. Douglas Hamelin, Thomas M. Shortell, and International Council on Systems Engineering, editors. *Systems engineering handbook: a guide for system life cycle processes and activities*. Wiley, Hoboken, New Jersey, 4th edition edition, 2015.
168. W.E. Walker, P. Harremoës, J. Rotmans, J.P. van der Sluijs, M.B.A. van Asselt, P. Janssen, and M.P. Krayer von Krauss. Defining uncertainty: A conceptual basis for uncertainty management in model-based decision support. *Integrated Assessment*, 4(1):5–17, 2003.
169. Peng Wang, Garry Robins, Philippa Pattison, and Emmanuel Lazega. Exponential random graph models for multilevel networks. *Social Networks*, 35(1):96–115, January 2013.
170. James Richard Wertz, David F. Everett, and Jeffery John Puschell, editors. *Space Mission Engineering: The New SMAD*. Microcosm Press, 2011.
171. Robert K Wysocki. *Effective project management: traditional, agile, extreme*. John Wiley & Sons, 2011.
172. Stanley Zionts and Jyrki Wallenius. An interactive programming method for solving the multiple criteria problem. *Management science*, 22(6):652–663, 1976.

Index

adjacency matrix, 89, 90
advanced air mobility, 206, 207
    stakeholders, 211
advanced air transportation system, 205
    abstraction phase, 211
    definition phase, 206
    implementation phase, 215
    problem formulation, 205
    resources, 208
    ROPE table, 208
agent-based modeling, 107
    ABM for SoS, 115
    air traffic management model, 117
    Behavior-Knowledge-Intention, 110
    constructs, 110
    disadvantages, 114
    Solar System Mobility Network, 117, 118
    types of agents, 111, 112
    why choose ABM?, 112, 114
air transportation system
    ROPE table, 209
analytic workbench, 121
    archetypal questions, 123
artificial intelligence, 185, 189
    patterns, 186
autonomy, 185

boundaries, 8

complex adaptive systems, 108
complex system, 9, 10
complexity, 193
    complex adaptive systems, 195
    definition, 195
    metrics, 197
        coupling complexity, 199
    quantification, 194
    sources, 196, 197
cybernetics, 6

design structure matrix, 182

emergent behavior, 10, 15, 33

finite state automata, 108

game of life, 109
general systems theory, 6

hierarchy
    complexity, 196
holism, 6
human space exploration, 221
    abstraction phase, 228
    architecture categories, 224
    definition phase, 222
    implementation phase, 232
    ROPE table, 223
    SDDA analysis, 235
    SODA analysis, 234
    technologies, 227
hypothesis, 64

INCOSE Systems Engineering Handbook, 40
ISO/IEC/IEEE 15288, 28
ISO/IEC/IEEE 21839, 13, 34
ISO/IEC/IEEE 21840, 34
ISO/IEC/IEEE 21841, 34

machine learning, 185
model-based systems engineering, 40, 199
    DoDAF, 200
    SysML, 200
multi-stakeholder dynamic optimization
    approximate dynamic programming, 178
    manager, 176
    participant, 177
    transfer contract, 175

network growth algorithm, 96
    Barabási-Albert, 97

random network, 97
scale-free network, 97
network measures, 89
assortativity, 95
betweenness centrality, 94
clustering coefficient, 93
degree distribution, 91
diameter, 93
eigenvector centrality, 94
network density, 91
node degree, 90
shortest path, 93
network theory, 85
cycle, 93
links, 86
nodes, 86
paths, 92
social network, 86
topology, 85
network topology, 88, 95
Random network, 88
Scale-free network, 88
network types
bipartite, 87
directed, 86
hypergraph, 87
undirected, 86
neural network, 189, 192

operations research, 6
optimization, 28
multidisciplinary optimization, 28

RAND Corporation, 3, 4
real-time strategy games, 187
game balance, 187
reductionism, 5
research question, 63
quantitative, 64
robust portfolio optimization
bertsimas-sim, 129
conditional value at risk, 131
constraints, 125
mean variance optimization, 126
system, 8
behaviors, 8
emergence, 9, 10
goals, 9, 16
objectives, 9
system dynamics, 7, 180
system of systems, 5, 13
abstraction, 39, 41
analytic workbench, 121
architecting heuristics, 31
distinguishing criteria, 14
history, 11
lexicon, 32, 41
ROPE Table, 41, 46, 47
stakeholders, 41, 45
taxonomy, 32, 43
tools, 121
types, 16
system of systems engineering, 29, 31
abstraction phase, 49
AI/ML, 186
DAI process, 39, 45
definition phase, 45
implementation phase, 53
mission engineering, 186
principles, 32
stakeholders, 32–34
verification and validation, 54
systems developmental dependency analysis
criticality of dependency, 161
dependencies, 159
deterministic analysis, 163
formulation, 161
stochastic analysis, 166
strength of dependency, 161
systems engineering, 23, 25
analysis, 28
definition, 23
modeling, 26
modeling practices, 27
requirements, 24, 30
Vee-model, 24, 30, 40
verification and validation, 24

Systems Engineering Body of Knowledge, 6, 34
systems operational dependency analysis
  criticality of dependency, 135
  dependencies, 133
  deterministic analysis, 142
  formulation
    multiple dependency, 137
    single dependency, 136
  impact of dependency, 135
  problem setup, 144
  resilience, 141
  robustness, 139
  stochastic analysis, 143
  strength of dependency, 135
systems thinking, 4, 5, 7, 23, 31, 32

uncertainty, 190
  uncertainty quantification, 192

wave model, 121

9781032138305